Mioara Petrus

Laser tissue – interaction applications

Mioara Petrus

Laser tissue – interaction applications

LAP LAMBERT Academic Publishing

Impressum / Imprint
Bibliografische Information der Deutschen Nationalbibliothek: Die Deutsche Nationalbibliothek verzeichnet diese Publikation in der Deutschen Nationalbibliografie; detaillierte bibliografische Daten sind im Internet über http://dnb.d-nb.de abrufbar.
Alle in diesem Buch genannten Marken und Produktnamen unterliegen warenzeichen-, marken- oder patentrechtlichem Schutz bzw. sind Warenzeichen oder eingetragene Warenzeichen der jeweiligen Inhaber. Die Wiedergabe von Marken, Produktnamen, Gebrauchsnamen, Handelsnamen, Warenbezeichnungen u.s.w. in diesem Werk berechtigt auch ohne besondere Kennzeichnung nicht zu der Annahme, dass solche Namen im Sinne der Warenzeichen- und Markenschutzgesetzgebung als frei zu betrachten wären und daher von jedermann benutzt werden dürften.

Bibliographic information published by the Deutsche Nationalbibliothek: The Deutsche Nationalbibliothek lists this publication in the Deutsche Nationalbibliografie; detailed bibliographic data are available in the Internet at http://dnb.d-nb.de.
Any brand names and product names mentioned in this book are subject to trademark, brand or patent protection and are trademarks or registered trademarks of their respective holders. The use of brand names, product names, common names, trade names, product descriptions etc. even without a particular marking in this works is in no way to be construed to mean that such names may be regarded as unrestricted in respect of trademark and brand protection legislation and could thus be used by anyone.

Coverbild / Cover image: www.ingimage.com

Verlag / Publisher:
LAP LAMBERT Academic Publishing
ist ein Imprint der / is a trademark of
OmniScriptum GmbH & Co. KG
Heinrich-Böcking-Str. 6-8, 66121 Saarbrücken, Deutschland / Germany
Email: info@lap-publishing.com

Herstellung: siehe letzte Seite /
Printed at: see last page
ISBN: 978-3-659-62800-9

Copyright © 2014 OmniScriptum GmbH & Co. KG
Alle Rechte vorbehalten. / All rights reserved. Saarbrücken 2014

Laser tissue – interaction applications:
Study of thermal effects, numerical simulation and quantitative analysis of surgical smoke

Edited by Dr. Mioara Petrus

National Institute for Laser, Plasma and Radiation Physic, Laser Department, 409 Atomistilor St., PO Box MG-36, 077125 Bucharest, Romania

Content

Preface	1
Chapter 1: Histophatology and Optical Coherence Tomography study of laser radiation interaction with tissue from the upper aerodigestive tract	3
1.1 Introduction	3
1.2 Laser-tissue: interaction mechanisms	4
1.3 Heat effects	7
1.4 Control of the surgical laser	9
1.5 Histological and Optical Coherence Tomography analysis of thermal effects in upper aerodigestive tract tissues using a CO_2 laser	10
1.5.1 Optical Coherence Tomography method	11
1.5.2 Materials and Methods	11
1.5.3 Results and discussions	12
1.6 Conclusions	17
References	17
Chaper 2: Numerical simulation of temperature distribution in laser irradiated tissue	20
2.1 Introduction	20
2.2 Modeling heat transfer in tissue	22
2.3 Mathematical model	26
2.4 Results and discussions	30
2.5 Conclusions	36
References	37
Chapter 3: Spectroscopic analysis of surgical smoke produced *in vitro* by laser vaporization of animal tissues in a closed gaseous environment	39
3.1 Introduction	39
3.2 Experimental section: Laser Photoacoustic Spectroscopy	40
3.3 Production and collection *in vitro* of surgical smoke samples	46
3.4 Results and discussions: quantitative analysis	47
3.5 Conclusions	50
References	50

Preface

Laser treatment based on controlled tissue elimination is well established as the treatment at the upper aerodigestive tract/ENT (Ear-Nose-Throat). When choosing a proper set of irradiation parameters (wavelength, pulse length, laser power, beam size, etc.) for a pulsing laser beam applied to a given target zone, some undesired kinds of tissue can be destroyed by inducing thermal damage in it, while the temperature of the surrounding tissues is kept below the threshold for damage. However the optimal choice of laser irradiation parameters and guidance of treatment is closely related to the prognosis of results. This is a very complex problem because it is strongly associated with the specific lesion characteristics from the upper aerodigestive tract, the surrounding tissue, the type of laser device with a diversity of laser irradiation parameters. The many issues involved in this problem made of it a field of actual interest that claims for an in-deep research. In the last, this problem has leaded a growing interest in modeling the interaction between laser irradiation and soft tissues from the upper aerodigestive tract.

We have address this problem by a new laser/tissue interaction model based on three different layers: (a) the thermal damage is predicted using Optical Coherence Tomography and histopathology analysis; (b) the temperature distribution in the tissue caused by laser energy deposition is estimated by solving the *bioheat* transfer equations, tissue models– is determined by Comsol MultiPhysics simulation; (c) quantitative analysis of surgical smoke produced after soft tissue ablation using CO_2 laser photoacoustic spectroscopy technique.

When any type of energy-based surgical instrument such as laser, high-frequency electric knives, and ultrasonic/harmonic scalpel is applied to human tissue, an unwanted by-product is produced, commonly known as surgical smoke. Aerosols, vapors and gases in the surgical smoke produced during laser vaporization of the tissue may constitute a threat to personnel and the patient health and can be a potential risk. The risk of surgical smoke is due to the odor, size of particles and gas concentration.

Laser photoacoustic spectroscopy (LPAS) has emerged over the last decade as a very powerful investigation technique, capable of measuring trace gas concentrations at ppmV (parts per million by volume), or even sub-ppbV (parts per billion by volume) level. LPAS provide several unique advantages, notably the multicomponent capability, high sensitivity and selectivity, wide dynamic range, immunity to electromagnetic interferences, convenient real time data analysis, operational simplicity, relative portability, relatively low cost per unit, easy calibration, and generally no need for sample preparation.

The present book focuses on the study of CO_2 laser – soft tissue interaction by analyzing: thermal effects produced in tissue, temperature distribution in tissue using Comsol MultiPhysics and surgical smoke produce by tissue vaporization. This book consists of 3 chapters; the first two chapters are focused on analyzing the thermal effects in soft tissue from the aerodigestive upper tract after CO_2 laser irradiation by Optical Coherence Tomography and histophatological, and temperature distribution in tissue using the bioheat equation. The third chapter is focused on quantitative analysis of a by product of laser – tissue interaction, surgical smoke using laser photoacoustic spectroscopy technique. The book represents a potentially interesting to researchers and specialists with the application of life sciences area.

About editor

Dr. Mioara Popa
Dr. Mioara Petrus followed the Faculty of Physics, University of Bucharest and has a PhD at the Faculty of Electronics, Telecommunications and Information Technology, University Polytechnic of Bucharest. Works as researcher at NILPRP-Department of Lasers, Romania and she has results in laser radiation – tissue interaction, numerical simulation of temperature distribution in irradiated tissue in Comsol MultiPhysics and laser photoacoustic spectroscopy applications.

Chapter 1
Histophatology and Optical Coherence Tomography study of laser radiation interaction with tissue from the upper aerodigestive tract

1.1 Introduction

Biomedical applications have gained a significant attention in recent years because of an increased interest in human health. From the beginning, lasers were considered as a very potential and promising tool for several different applications in medicine [1-13]. Nowadays, the use of laser has become a very common practice in all the medicine sectors and with different pathologies. These pathologies have been cured thanks to the use of different laser sources, selected as a function of wavelengths and power, and to more and more sophisticated technologies. The heat from the laser beam irradiation exposed tissue may lead to vaporization, thermal necrosis of the adjacent tissue, charring and release of surgical smoke and water vapors. The carbonization of cells that creates surgical smoke affects medical personnel and patients by launches in the air harmful chemical and biological particles. Due to different thermal and optical properties of biological tissues, laser beam parameters, complexity of physical and biochemical process involved, it is difficult to reach a set of experiments test to solve medical problems. That is why a special attention is accorded to modeling and numerical simulation. There are several mathematical approaches describing the laser radiation interaction with biological tissue: first order scattering, Kulbenka-Munk theory, Monte Carlo simulation, but a special attention was accorded to the transfer heat equation [2,8,10-12].

CO_2 laser produce light with a wavelength of 10.6 µm in the infrared range of the electromagnetic spectrum. The radiant energy produced by the CO_2 laser is strongly absorbed by pure, homogeneous water and by all biological tissues high in water content [1,5,13]. The extinction length of this wavelength is about 0.03 mm in water and in soft tissue; reflection and scattering are negligible. Because absorption of the radiant energy produced by the CO_2 laser is independent of tissue color and because the thermal effects produced by this wavelength on adjacent tissues are minimal, the CO_2 laser has become extremely versatile for use in otolaryngology surgery. With current technology, light from this laser cannot be transmitted through existing flexible fiberoptic endoscopes, although research and development of a suitable flexible fiber for transmission of this wavelength is being carried out on an international level. Presently, the radiant energy of this laser is transmitted from the optical resonating chamber to the target tissue via a series of mirrors through an articulating arm to the target tissue (Ossoff and Karlan, 1985). This laser can be used free-hand for macroscopic surgery, attached to the operating microscope for microscopic surgery, and adapted to an endoscopic coupler for bronchoscopic surgery

(Ossoff and Karlan, 1982; Ossoff et al, 1985); in this application, rigid, nonfiberoptic bronchoscopes must be used (Ossoff and Karlan, 1983b).

The CO_2 laser creates a characteristic wound. When the target absorbs a specific amount of radiant energy to raise its temperature to between 60°C and 65°C, protein denaturation occurs. Blanching of the tissue surface is readily visible and the deep structural integrity of the tissue is disturbed. When the absorbed laser light heats the tissue to approximately 100°C, vaporization of intracellular water occurs. This causes vacuole formation, crater, and tissue shrinkage. Carbonization, disintegration, smoke, and gas generation with destruction of the laser radiated tissue occurs at several hundred degrees centigrade. In the center of the wound is an area of tissue vaporization; here just a few flakes of carbon debris are noted. Immediately adjacent to this area is a zone of thermal necrosis measuring approximately 100 microm wide. Next is an area of thermal conductivity and repair, usually 300 to 500 microm wide. Small vessels, nerves, and lymphatics are sealed in the zone of thermal necrosis; the minimal operative trauma combined with the vascular seal probably account for the notable absence of postoperative edema characteristic of laser wounds (Mihashi et al, 1976).

Otorhinolaryngology is concerned with diseases of the ear, the nose, and the throat (ENT). Many procedures in the upper aerodigestive tract that previously required prolonged hospitalization and tracheotomy can now be performed without the need for tracheotomy (Hollinger, 1982), and often as an outpatient procedure. It has already been shown by Jako (1972) and Strong et al. (1976) that benign and malignant lesions of the glottis, i.e. the vocal cords, can be more safely removed with the CO_2 laser rather than mechanically. Major indications for laser treatment of the nose are highly vascularized tumors such as hemangiomas or premalignant alterations of the mucosa. The principal advantage of lasers is again their ability to simultaneously perform surgery and coagulate blood vessels. Six types of lasers are commonly in use in otolaryngology - head and neck surgery, and many more are in various stages of development. The six types of lasers in use are the argon laser, the argon tunable dye laser, Nd:YAG laser, KTP laser, flash lamp pumped dye laser, and CO_2 laser. The potential clinical applications of each of these surgical lasers are determined by their wavelength and specific tissue absorptive characteristics. Therefore the surgeon should consider the properties of each wavelength when choosing a particular laser. This will facilitate the achievement of his surgical objective with minimal morbidity and maximal efficiency.

1.2 Laser – tissue: interaction mechanisms

When electromagnetic energy (incident radiation) interacts with tissue, the tissue reflects part, the tissue absorbs part, and the tissue transmits and scatters part of the light. The surgical interaction of this radiant energy with tissue is caused only by that portion of the light that is absorbed (Polanyi, 1983) (see Fig. 1.1). The tissue effects produced by the radiant energy of a laser vary with the specific wavelength of the laser used. Each type of laser exhibits characteristic and different biological effects on tissue and is therefore useful for different applications. Specific laser characteristics as well as laser parameters contribute to this diversity. Most important among optical tissue properties are the coefficients of absorption, reflection, scattering and others such as heat conduction and heat capacity. All together, they determine the total transmission of the tissue at certain wavelengths.

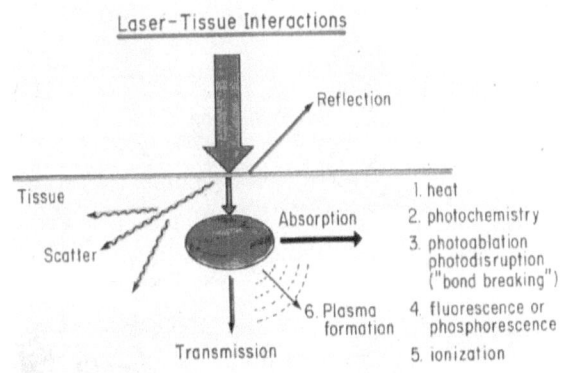

Fig. 1.1 The mechanisms in laser-tissue interaction

There are several mechanisms by which laser radiation interact with biological tissue. Interaction mechanisms used in medical and therapeutic application are [8]:
1. Photochemical reactions: when a molecule absorbs a photon, the energy in the photon is transferred to the molecule's electrons. More energetic electrons can more easily escape the nuclear forces keeping them close to the nuclei, and so excited molecules are more likely to undergo chemical reactions (exchanging or sharing of electrons) with other molecules. In photodynamic therapy, for instance, a photosensitising drug (which consists of molecules that become very reactive when they absorb light) causes reactive oxygen species to form, which leads to both necrosis (cell death) and apoptosis ('programmed' cell death). Photodynamic therapy is increasingly widely used in oncology to destroy cancerous tumours.
2. In photothermal interactions, the energy of the photons absorbed by chromophores (a name given to any light-absorbing molecules) is converted into heat energy, which can cause a range of thermal effects from tissue coagulation to vaporization. Applications include tissue cutting and welding in laser surgery.
3. In photoablation, high-energy, ultraviolet (UV) photons are absorbed and, because they are more energetic than the chemical bonds holding the molecules together, cause the dissociation of the molecules. This is followed by rapid expansion of the irradiated volume and ejection of the tissue from the surface. This is used in eye (corneal) surgery, among other applications.
4. In plasma-induced photoablation a free electron is accelerated by the intense electric field in the vicinity of a focused laser beam. When this very energetic electron collides with a molecule, it gives up some of its energy to the molecule. When

sufficient energy is transferred, a bound electron is freed, and a chain reaction of similar collisions is initiated, resulting in plasma: a soup of ions and free electrons. One application of this is in lens capsulotomy to treat secondary cataracts.

5. The final set of related mechanisms, grouped under the term photodisruption, is the mechanical effects that can accompany plasma generation, such as bubble formation, cavitation, jetting and shockwaves. These can be used in lithotripsy (breaking up kidney or gall stones), for example.

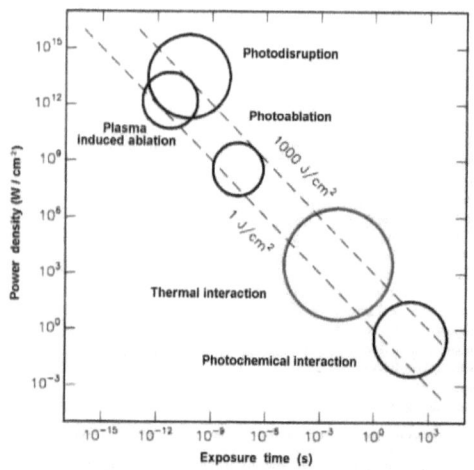

Fig. 1.2 Map of laser-tissue interaction [8]

In Fig. 1.2 the y axis expresses the applied power density in W/cm² and the exposure time in seconds is reported on the x axis. According to this map, the time scale can be roughly divided into five sections: continuous wave or exposure times more than 1 s for photochemical interaction, from 1 s to 1 μs for thermal interaction, from 1 μs 1 ns for photoablation, and less than 1 ns for plasma-induced ablation and photodisruption. The difference between the latter two is attributed to different energy densities [8].

All these mechanisms of interaction will depend on:
1. type of molecules from biological tissue;
2. wavelength of laser radiation;
3. the intensity of the laser beam;
4. exposure time.

For the energy in the photons to end up as heat in tissue, two things must happen [8]:

1) Absorption: The photon must be absorbed by a molecule, putting that molecule into an excited state. When we are looking at soft tissue and the absorption of infrared light, as we will be in this section, the main chromophores (light absorbers) of interest will be water molecules.

2) Vibrational relaxation: Collisions with other molecules leads to a gradual deactivation of the original molecule (the excited electron moves down the ladder of permitted energy levels) and an increase in the kinetic energy of those it collides with.

An increase in the kinetic energies of the molecules is, on a macroscale, an increase in the temperature of the tissue.

With photothermal effects, there is no specific pathway, and the photons may be absorbed by any biomolecule and still lead to a thermal effect (see Fig. 1.3). Heat energy is deposited in the tissue by the absorption of light and its subsequent conversion to heat via vibrational relaxation. This causes a rise in temperature of the tissue. Then the heat will diffuse through the tissue causing a rise in temperature in the surrounding tissue. The damage done to the tissue depends on the temperature that is reached, and the duration at which it is held at that temperature. There are many different and varied medical applications that use a thermal interaction, from vaporization of tumors, to welding gastrointestinal ulcers, and the removal of skin marks such as port wine stain birthmarks or tattoos.

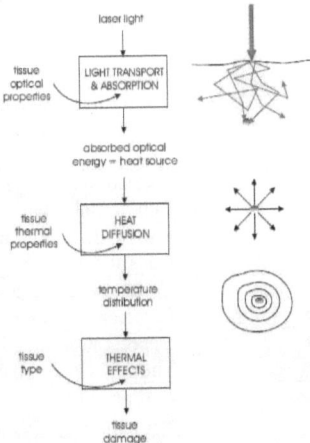

Fig. 1.3 The various aspects involved in thermal interaction of light with tissue [8].

1.3 Heat effects

The first mechanism by which tissue is thermally affected can be attributed to conformational changes of molecules [1,2,8]. These effects, accompanied by bond destruction and membrane alterations, are summarized in the single term hyperthermia ranging from approximately 42-50 °C. If such a hyperthermia lasts for several minutes, a significant percentage of the tissue will undergo necrosis as described below by Arrhenius' equation. Beyond 50°C, a measurable reduction in enzyme activity is observed, resulting in reduced energy transfer within the cell and immobility of the cell. Furthermore, certain repair mechanism of the cell is disabled. Thereby, the fraction of surviving cells is further reduced. At 60°C, denaturation of proteins and collagen occurs which leads to coagulation of tissue and necrosis of cells. The corresponding macroscopic response is the visible paling of the tissue. At even higher temperatures more then 80 °C the membrane permeability is drastically increased, thereby destroying the otherwise maintained equilibrium of chemical concentrations. At 100°C, water molecules contained in most tissues start to vaporize. The large vaporization heat of water (2253 KJ/Kg) is advantageous, since the vapor generated carries away excess heat and helps to prevent any further increase in the temperature of adjacent tissues. Due to the large increase in volume during this phase transition, gas bubbles are formed inducing mechanical ruptures and thermal decomposition of tissue fragments. Only if all water molecules have been vaporized, and laser exposure is still continuing, does the increase in temperature proceeds. At temperatures exceeding 100°C, carbonization takes place which is observable by the blackening of adjacent tissues and the escape of smoke. To avoid carbonization, the tissue is usually cooled down with either water or gas. Finally, beyond 300°C, melting can occur, depending on the target material. All these steps are summarized in table 1.1, where the local temperature and the associated tissue effects are listed.

Tabel 1.1 Thermal effects of laser radiation with biological tissue

Temperature (^{0}C)	Biological effects
43-45	- Conformational alteration - Shrinkage of the tissue - Hyperthermia
45-60	- Alteration of enzymatic activity - Cell immobility
60-75	- Protein denaturation - Coagulation
75-100	- Collagen denaturation - Permeabilization of membranes
100	- Vaporization - Thermal decomposition (ablation)
>100	- Carbonization - Formation and breaking of vacuols
>300	- Thermoablation of the tissue

Thermal effects are perhaps the most widely encountered form of tissue-laser interaction in clinical practice (see. Fig. 1.4). Frequently, not only one but several thermal effects are induced in biological tissues, depending on the laser parameters. These effects might even range from carbonization at the tissue surface to hyperthermia a few millimeters inside the tissue. Since the critical temperature for cell necrosis is determined by the exposure time, no well-defined temperature can be declared which distinguishes reversible from irreversible effects. Thus, exposure energy, exposure volume, and exposure duration all together determine the degree and extent of tissue damages.

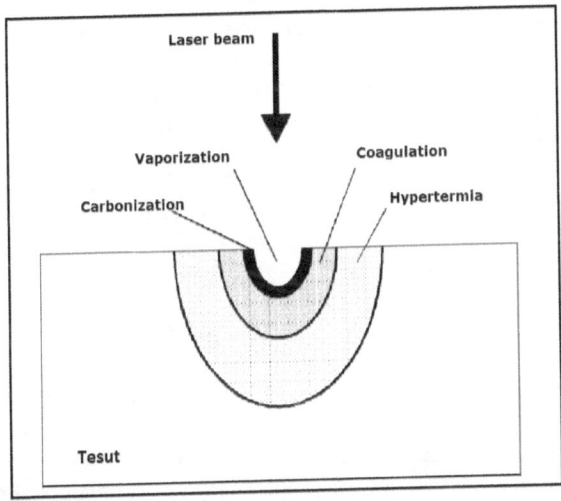

Fig. 1.4 Thermal effects on tissue al laser radiation interaction

1.4 Control of the surgical laser

With most surgical lasers, the physician can control three variables: (1) power (measured in watts), (2) spot size (measured in millimeters), and (3) exposure time (measured in seconds). Of these three variables, power is the least useful as a parameter and may be kept constant with widely varying effects, depending on the spot size and the duration of exposure. For example, the relationship between power and depth of tissue injury becomes logarithmic when the power and exposure time are kept constant and the spot size is varied (Ossoff and Karlan, 1985).

Power density (PD) is a more useful measure of the intensity of the beam at the focal spot than power because it takes into account the surface area of the focal spot. Specifically, power density or power per unit area of the beam, expressed in watts per square centimeter, is a measure of the power output of the laser in watts divided by the cross-sectional area of the focal spot in square centimeters [1].

PD = (Power in the focal spot) / (Area of the focal spot).

Power and spot diameter are considered together and a combination is selected to produce the appropriate power density. If the time of exposure is kept constant, the relationship between power density and depth of injury is linear as the spot size is varied. Power density is the most important operating parameter of a surgical laser at a given wavelength. Therefore surgeons should calculate the appropriate power density for each procedure to be performed; these calculations would allow the surgeon to control in a predictable manner the tissue effects when changing from one focal length to another (400 mm for microlaryngeal surgery to 125 mm for hand-held surgery) or when using surgical lasers with different transverse electromagnetic modes (TEM00 versus TEM01). Power density varies directly with power and inversely with surface area (A). This relationship of surface area to beam diameter is important when evaluating the power density. The larger the surface area, the lower the power density; conversely, the smaller the surface area, the higher the power density. Surface area is expressed as:

$$A = pir^2$$

where, r is the beam radius. Because the radius is one-half of the beam diameter ($d/2$), surface area also can be expressed as:

$$A = (pid^2) / 2^2 \text{ or } A = (pid2) / 4.$$

Newer CO2 lasers emit radiant energy with a characteristic beam intensity pattern different from that produced by older-model lasers. Because this beam pattern ultimately determines the depth of tissue injury and vaporization pattern across the focal spot, the surgeon must be aware of the characteristic beam pattern of the laser. *Transverse electromagnetic mode* (TEM) refers to the distribution of energy across the focal spot and determines the shape of the laser's spot. The most fundamental transverse electromagnetic mode is TEM_{00}, appearing circular when cut in cross section; the power density of the beam follows a Gaussian distribution, with its greatest amount of energy at the center of the beam, then diminishing progressively toward the periphery. TEM_{01} and $TEM1_1$ modes are less fundamental modes that have a more complex distribution of energy across their focal spot, causing predictable variations in tissue vaporization depth. Additionally, their beams cannot be focused down to as small a spot size at the same working distance as TEM_{00} lasers (Fuller, 1980).

1.5 Histological and Optical Coherence Tomography analysis of thermal effects in upper aerodigestive tract tissues using a CO_2 laser

Lasers have been widely considered in clinical applications. The widespread use of infrared (IR) lasers for medical applications is based on the fact that water, the constituent of soft l tissues, strongly absorbs optical radiation in the IR region. One of the most important lasers in IR that widely used today is the carbon dioxide (CO_2) laser. Since the targeted chromophore is water, the CO_2 laser is commonly used, because of its relatively strong absorption by water in the far-infrared wavelength range (Eze and Kumar, 2010). The CO_2 laser only penetrates superficially owing to the high absorption coefficient of tissue water that allows minimal thermal damage and enough to cause tissue vaporization. The dermatology applications such as skin resurfacing and tissue removal made use of these featured [2-4].

In the recent years, laser technology has been introduced as an integral part of the therapeutic means in many centers of otolaryngology, head and neck surgery [1,2,20-23]. In laser surgery pulse laser radiation is used to ablate the tissue and the goal is to minimize the collateral thermal damage in a desired region. The tissue at the aerodigestive tract is mainly formed by soft tissue, and is necessary a very precise cut, with a minimum thermal damage in the surrounded tissue. It has already been shown that benign and malign lesions of the glottis, the vocal cords, can be safely removed with the CO_2 laser rather than mechanically [1]. The vocal folds of the larynx permit the capacity for phonation. Structural changes on vocal folds affect the performance of phonation. Due to the expansion of lesions into deeper layers or tissue loss, the voice becomes unstable and will be limited in frequency and dynamic range. Moreover, scarring of the tissue caused by inadequate wound healing or disregard for the layered structure during surgery can implicate permanent disphonia [1].

To characterize the (ear-nose-throat) ENT soft tissue destruction after CO_2 laser ablation we use the optical coherence tomography (OCT) and histological analysis. OCT and histological estimates of the ablation crater dimensions and the depth of thermal injury were obtained. The CO_2 laser is the most optical source used in the ENT surgery, with a wavelength of 10.6 µm may be used for precisely cutting or vaporizing soft tissue with hemostasis. This long wavelength is strongly absorbed by water and has a penetration depth of about 10 µm into the tissue. Compared to the CO_2 laser, the application range of the other laser systems (diode laser, Er:YAG, Nd:YAG, Ho:YAG) is limited due to the specific characteristics of each type of laser radiation, such as limited cutting quality or the selective absorption of the laser light by pigments, especially hemoglobin, melanin and myoglobin. The main constituents of biological tissue which contribute towards absorption in the near infrared are water, fat and hemoglobin [1,2,17].

The objective of this study is to estimate the interaction effects of a CO_2 laser in soft tissue, the crater ablation depth and width (in relation with power and pulse duration) and to investigate experimentally and theoretically the thermal response of soft tissue irradiated by a laser source.

1.5.1 Optical Coherence Tomography method

Optical Coherence Tomography (OCT) is a novel non-invasive imaging technique based on low-coherence interferometry well established in ophthalmology and dermatology [14-16]. It is also well suited for imaging of the microstructures of the vocal cords. OCT technique is capable of producing cross-sectional images with an axial resolution of less than 10 µm and a depth penetration of 1 to 2 mm in biological tissue. High sensitivity, large dynamic range, and micrometer level resolution imaging are achieved with an OCT technique by interferometric detection of backscattered light from the sample [6, 7, 8]. The OCT system used in our study is based on Fourier- Domain [9], Thorlabs OCP930SR Spectral Radar OCT device. The base unit contains a broadband super luminescent diode (SLD), a 930 nm light source, of 2 mW optical power, that is direct through an optical fiber couplet to the sample via the handheld scanning probe. The scanning device has a spectral bandwidth of 100 nm, 20 µm lateral resolution, 6.2 µm axial resolution and an image depth of approximately 1.6 mm. Images were acquires at 8 frames per second, an image size of 512 rows and the duration of the scan acquisition was 5 to 10 seconds.

1.5.2 Material and methods

The interaction process of tissue from the upper aero digestive tract (such as vocal cords) in the presence of laser radiations was investigated histological and with an optical tomography. As biological materials we used *in vitro* samples of tonsil and porcine vocal cords. The irradiation of tissue samples was made by a laser system, a CO_2 laser surgical system (SP-25LA, China), in pulse regime, the maximum output power for laser system being 25 W. For a good positioning precision and reproducibility, the laser radiation is applied on vocal cords with a micromanipulator (Acuspot 711 Sharplan) under microscopic control and the laser beam had a Gaussian profile. For the irradiation was made two sets of tissue samples, one for the OCT analysis and, one was sent to the histological laboratory for the comparative analysis. The two sets of samples were made in the same condition and were irradiated with the same laser parameters: power, pulse duration and beam shape. After the irradiation, one set of sample was analyses with the OCT (Spectral Radar Coherence Tomography –SROCT ThorLab version) and we obtained images with the ablation crater, i.e., width and depth of the laser ablation crater. The refractive index of porcine vocal cords, for the OCT measurements was $n = 1.39$. Each OCT image is an optical cross-section of the tissue sample, and was obtained with a time delay not larger than one hour from the irradiation process. Optimal histological analyze requires a procedure which can take few days. This can cause modifications in the sample structure. For histological analysis, the tissue was fixed in formaldehyde, serially dehydrated in graded ethanol baths, embedded in paraffin, and sectioned in micro sections with thickness of ~ 6 μm. Digital images were obtained with an optical microscope.

1.5.3 Results and discussion

In a first analysis of laser radiation effects in biological tissue, we tested a CO_2 surgical laser system on tonsil tissue. As is known, the nasal mucosa is a tissue is a tissue with a multiple structure, muscle, salivary gland, blood vessels, and therefore the interaction with laser radiation is a complex phenomenon. Tonsil tissue samples were excised by classic surgical procedure from a patient (a patient of 16 years). The tissue was irradiated in vitro using a CO_2 surgical system (CO2 Surgical Laser System, Model: SP25-LA) immediately after surgical procedure. The tissue sample was irradiated at different laser power and exposure time in order to determine the optimum conditions for surgical procedures. Tonsillar tissue is composed mainly of lymphoid follicles with different structural composition, muscular structure, blood vessels and lymphoid tissue. Due to this variation in composition is necessary to control laser parameters.

After laser irradiation, tissue samples were processed in the histophatological laboratory to the standard procedure. This procedure requires the preservation and placing the sample in buffer solution. Then, the sample are processed with paraffin and cut into very thin section of 3-5 microns. Pathologist analyzes the sample using an optical microscope. In Fig. 1.5-1.9 are shown the images obtained using an optical microscope after the histophatological analysis of tonsil samples after CO_2 laser ablation at different laser powers (P = 10 W, 15 W, 25 W) and an exposure time of 100 ms and 500 ms.

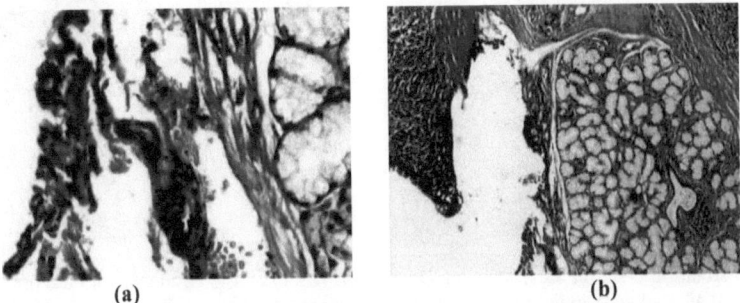

Fig. 1.5 Thermal necrosis in tonsil after CO_2 laser ablation at P = 25 W, t_{exp}= 500 ms: a) necrosis and coagulation area in tissue, red blood cells with a size of about 7 μm, and the tissue edema observed with a presence of lymphatic vessels; b) view of the crater with the presence of intact red blood cells in the salivary gland.

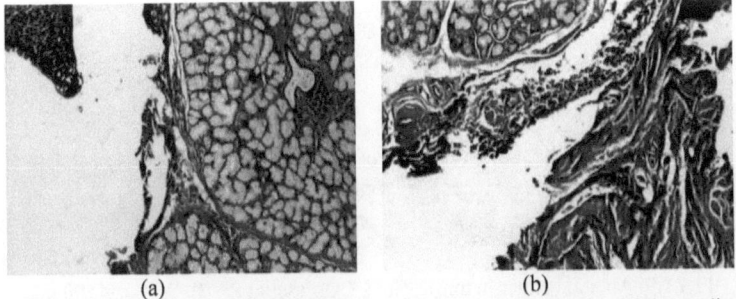

Fig. 1.6 Thermal necrosis in tonsil after CO_2 laser ablation: (a) crater ablation; right – salivary gland , left - muscle, P = 25W, t_{exp}= 500 ms; (b) crater ablation with diameter ~ 800 μm, P = 25W, t_{exp}= 500 ms

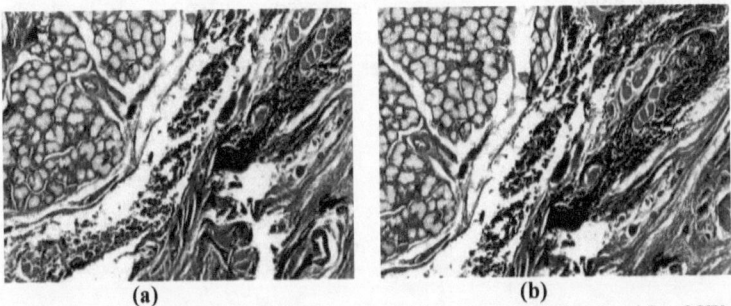

Fig 1.7 Thermal necrosis in tonsil after CO_2 laser ablation at t_{exp} = 500 ms and P = 25W: (a) tip ablation crater and thermal affected tissue due to heat propagation after a long exposure time of 500 ms; (b) adjacent thermal affected zone in the tip of crater ablation

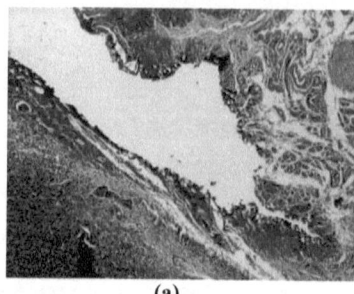 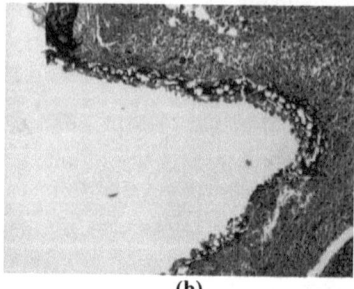

Fig. 1.8 Ablation crater in tonsil with a CO_2 laser (a) P = 15W, t_{exp}= 500 ms, with the presence of the thermal afected zone in the tip of the crater after a long exposure of 500 ms; (b) Ablation crater at P = 25W, t_{exp} = 100ms, with the presence of a small necrosis zone.

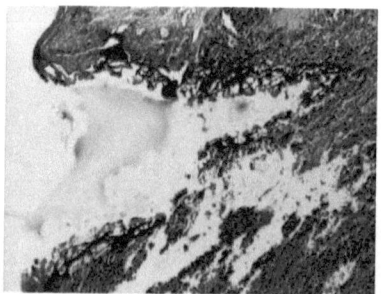

Fig. 1.9: Ablation crater in tonsil with a CO_2 laser at P = 10 W, t_{exp} = 500 ms; can be seen a large area of tissue affectde by the heat

In Fig. 1.10 can be seen a comparison of images obtained with an optical microscope (hisotpatological analysis) and confocal microscope after a CO_2 laser ablation of upper aerodigestive tract tissue at laser parameters of P = 25 W and exposure time of 100 ms. The crater ablation has a diameter of 850 µm and a depth of 670 µm and an area of necrosis between 55 and 70 µm.

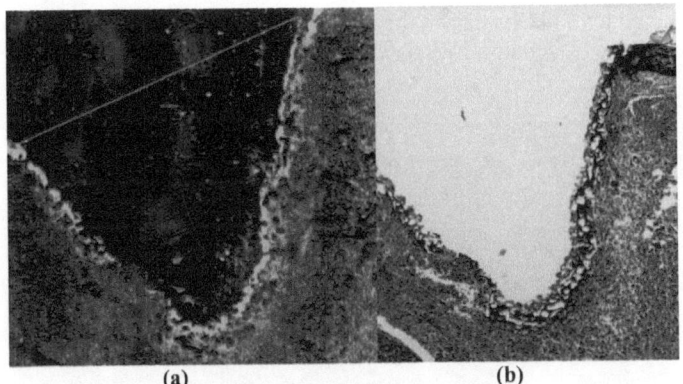

Fig. 1.10 Crater ablation in amygdala with a CO_2 at P = 25W si t_{exp}= 100 ms ; (a)confocal microscope; (b) optic microscope – histological analysis

In the OCT images, the high reflectivity at the burn surfaces is due to carbonization zone. OCT and histology showed no significant differences, the crater depth is directly proportional to laser power, the ablation crater has a conic shape, with an input aperture of about 800 μm diameter and a depth of ~ 1 mm, which is in accordance with the focal length of the laser beam focusing system (f = 125 mm) and with the laser beam divergence. The crater walls presents necrosis with a thickness of 60-80 μm (the necrosis thickness was evaluated taking into account the diameter (6-7 μm) of the hemoglobin cells and lymphocytes cells visible in the image of the optical microscope and in the OCT images we measure the crater ablation dimensions using the corresponding software.

Fig. 1.11 show the porcine vocal cords after CO_2 laser ablation at different laser power and exposure time of 100 ms and comparison of thermal effects with OCT and after histopathological analysis. Fig. 1.11 from left side shows the histological images obtained at different laser beam powers with the same exposed time i.e., t = 100 ms. As one can see the ablation volume and the thermal damage in the surrounding tissue increase with the power of the laser beam. At the lower power of the laser beam i.e., the 6 W, the irradiation effects are observed only as a necrosis zone there is no ablation crater in the tissue (see Fig.1.11 a-b).

By increasing with more than a factor two the power of the laser beam, i.e., 14 W, a smooth crater in a zone with muscle from the vocal cords is observed (see Fig. 1.11c - d). For this sample there is no zone of necrosis because it was past a long time from the moment of the vocal cords porcine extraction. At 16 W of laser beam power, one can see the crater ablation, with a small zone of necrosis on the edge, the lymphatic system is intact, and there is no damage caused by the heat propagation (see Fig.1.11 e –f). At high power, above 20 W, the histological images show a clear crater ablation with small necrosis zone and with visible hemoglobin and lymphocytes cells (see Fig.1.10 g –h obtained for 20 W and Fig. 1.11 i - j for 24 W). Figures from right side shows the OCT image obtained at different power laser beam with the same exposed time i.e., t = 100 ms, i.e., the same experimental parameters used for the tissue irradiation with the laser beam, as those used in histological analysis. The irradiation area is indicated by a bleak ring on the figures. The image observed at small distance, below approximately 1.5 mm, is due to the junction of the vocal cords.

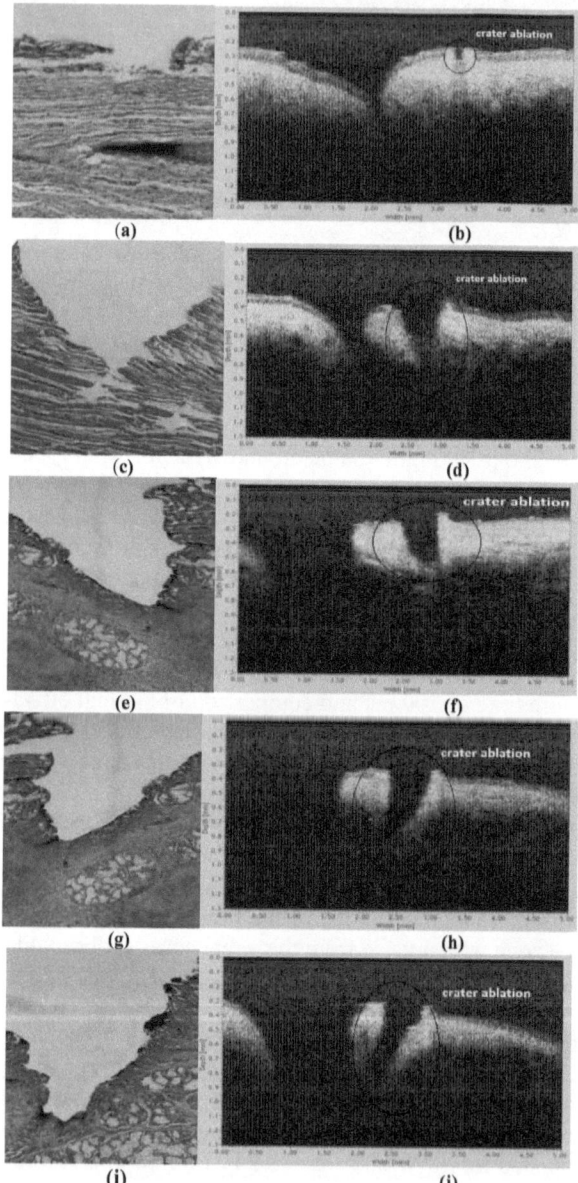

Fig. 1.11 Porcine vocal cords ablation by CO_2 laser at t = 100 ms and P = 6 W, 14 W, 20 W and 24 W: (a), (c), (e), (g), (i) - histological images, (b), (d), (f), (h), (j) - OCT image

1.6 Conclusions

In this study, as biological samples were used excised human tonsils by classic surgical procedure and pig vocal cords. Samples were irradiated using a CO_2 laser at different laser powers and different exposure time to determine the optimal laser parameters in surgical procedures. After irradiation tissue samples were analyzed by the histophatology and using an OCT system.

The exposure time is a very important parameters, a high exposure time determine thermal effects in adjacent tissue with the presence of a large area of thermal damage on the top of laser ablation crater.

In order to obtain a small area of thermal damage, but also to prevent the spread of heat in the tip of the crater ablation is required a short exposure time even at a high laser power. Therefore, to obtain a smooth crater ablation with a small area of necrosis and a minimal thermal damage it is necessary a small exposure time, to control the laser power and laser exposure time.

Comparison between the optical coherence tomography and conventional histological analysis demonstrate a very good agreement in the evaluation of the CO_2 laser injury of porcine larynges. The grate advantages offered by the optical coherence tomography can be used in the monitoring of the CO_2 laser surgery procedures.

A very simple model was used to describe de temperature distribution in the soft tissue. The numerical results obtained are in good agreement with the optical coherence tomography and conventional histological experimental data. Moreover, given the experimental parameter the threshold value for the ablation process of the CO_2 laser beam power was found.

The OCT can be an ideal method for real-time monitoring of vocal fold microsurgery due the adequate imaging depth and its high resolution and the surgeon could realize the therapeutic protocol using the simulation methods for the wanted result.

References

1. *R.H. Ossoff, L. Reinisch*, Chapter 12: Laser surgery: Basic Pronciples and safety considerations, Cummings otolaryngology: head and neck surgery, Philadelphia: Elsevier Mosby, pp. 214-230 (2005).
2. *M. H. Niemz*, Laser-tissue interactions, Fundamentals and Applications, Chapter 2- Light and Matter, 2,5 Photon Transport Theory, Third Enlarged Edition, Springer (2007).
3. *Seiner, Laser and IPL Technology in Dermatology and Aesthetic Medicine, Cap: Laser-tissue interactions,* Springer-Verlag Berlin Heidelberg (2011).
4. *G. Muller, B. Schaldach*, Basic laser tissue interaction, Lasers in medical science, Supplement 4(1):7 (1989).
5. *M. Maurizi, G. Almadori, G. Plaudetti*, Laser carbon dioxide cordectomy versus open surgery in the treatment of glottic carcinoma: our results, Otolaryngol. Head Neck Surg, 132 (2005).

6. *K. Goharkhay, A. Moritz, P. Wilder-Smith, U. Schoop, W. Kulger, S.Jacolitsch, W. Sperr*, Effects on oral soft tissue produced by a diode laser in vitro, Lasers in Surgery and medicine 25:401-406 (1999).
7. *J. Russ*, The laser radiation and biological medium, CRC Press (1995).
8. *B. Cox*, Optics in Medicine. Introduction to Laser-Tissue Interactions, Photoacoustic Imaging Group, Department of Medical Physics and Bioengeneering, University College London (2010), (www.medphys.ucl.ac.uk/~bencox/laser-tissue.pdf).
9. *B. Choi, A.J. Welch*, Analysis of thermal relaxation during laser irradiation of tissue, Lasers in Surgery and Medicine, 29 (4): 351-359 (2001).
10. *P.R. Sharma, S. Ali, V.K. Katiyar,* Transient heat transfer analysis on skin surface and inside biological tissue, Int. J. of Appl. Math and Mech 5(5):36-47 (2009).
11. *M. Bachmann*, Monte Carlo Simulations, Cornell University Library, (2011).
12. *H.H. Pennes*, Analysis of tissue and arterial blood temperatures in the resting human forearm, Journal of Applied Physiology 1: 93–122 (1948).
13. *V. Oswal, M. Remacle, S. Jovanovic, J. Krespi*, Principle and Practice of Lasers in Otolaryngology and Head & Neck Surgery, Kulger Publications (2002).
14. *W. Drexler, J.G. Fujimoto*, Optical Coherence Tomography, Technology and Applications, Biological and Medical Physics, Biomedical Engineering, ISSN 1618-7210, ISBN 978-3-540-77549-2, Springer (2008).
15. *K. Lüerßen, H. Lubatschowski, N. Radicke, M. Ptok*, Optical characterization of vocal folds using Optical Coherence Tomography, Medical Laser Application, 21:185-190 (2006).
16. http://www.thorlabs.com/catalogpages/595.pdf
17. *R. Leitgeb, W. Drexler, A. Unterhuber, B. Hermann, T. Bajraszewski, T. Le, A. Stingl, A. Fercher*, Ultrahigh resolution Fourier domain optical coherence tomography, Optics Express, 12(10): 2156 -2165 (2004).
18. *A.M. Forsea, E.M. Carstea, L. Ghervase, C. Giurcaneanu, G. Pavelescu*, Clinical application of optical coherence tomography for the imaging of non-melanocytic cutaneous tumors: a pilot multi-modal study, Med. Life, 3(4): 381-389 (2010).
19. *G.J. Jako*, Laser surgery of the vocal cords; an experimental study with carbon dioxide lasers on dogs, Laryngoscope, 82: 2204-2216 (1972).
20. *B.A.Torkian, S. Guo, A.W. Jahng, L-H L. Liaw, Z. Chen, B.J.F. Wong*, Noninvasive measurement of ablation crater size and thermal injury after CO_2 laser in the vocal cord with Optical Coherence Tomography, Otolaryngology-Head and Neck Surgery, 134: 96-91 (2006).
21. *H. Wisweh, U. Merkel, A.K. Huller, K. Luerben, H. Lubatschowski*, Optical coherence tommography of vocal fold femtosecond laser microsurgery, Therapeutic Laser Applications and Laser-Tissue Interactions III, Proc. SPIE, 6632:527-533 (2007).
22. *M. Petrus, E.M. Carstea, D. C. A. Dutu, D. C. Dumitras*, Optical coherence tomography monitoring of laser ablation processes in ENT tissues, Journal of Optoelectronics and Advanced Materials, 13 (7): 776 – 780 (2011).

23. *M. Petrus, D.C. Dumitras*, Numerical simulation and *in vitro* experimental temperature distribution analysis in irradiated tissue, U.P.B. Sci. Bull. A, 74 (4): (2012).

Chaper 2
Numerical simulation of temperature distribution in laser irradiated tissue

2.1 Introduction

The extensive use of laser as a surgical tool has lead to a growing interest in modeling the interactions between laser radiation and biological tissues [1-3]. The thermal effects can be analyzed by OCT, which is a novel non-invasive imaging technique based on low-coherence interferometry well established in ophthalmology and vocal cords. OCT technique is capable of producing cross-sectional images with an axial resolution of less than 10 μm and a depth penetration of 1 to 2 mm in biological tissue. High sensitivity, large dynamic range, and micrometer level resolution imaging are achieved with an OCT technique by interferometric detection of backscattered light from the sample.

The wavelength of 10.6 μm is strongly absorbed by the water and the CO_2 laser is suitable for vaporization and the precise thermal cutting of tissue and can be considered as a device whose irradiation is absorbed near the surface [1-3]. The deposition of laser energy in the tissue is not only a function of laser parameters such as wavelength, power density, exposure time, spot size, and repetition rate. It also depends on optical tissue properties like absorption and scattering coefficients and on the opto - thermal tissue parameters. Temperature is the governing parameter of all thermal laser– tissue interactions. And, for the purpose of predicting the thermal response, a model for the temperature distribution inside the tissue must be derived. Development of a mathematical model for thermal injury must include determination of the light distribution, rate of heat generation, and heat transfer. One of the more used mathematical models for describing the temperature distribution inside biological tissues is the Pennes bioheat equation [5-7].

Heat generation is determined by laser parameters, *i.e.,* irradiance, exposure time and the absorption coefficient (who is a function of the laser wavelength), and optical tissue properties [1,2]. Heat transport is characterized by thermal tissue properties such as heat conductivity and heat capacity. Heat effects depend on the type of tissue and the temperature achieved inside the tissue. According to the degree of heating, selective thermal damage can be achieved: 42–45 °C: beginning of hyperthermia, conformational changes, and shrinkage of collagen; around 50 °C: reduction of enzymatic activity; around 60 °C: denaturation of proteins, coagulation of the collagens, membrane permeability; around 100 °C: tissue drying and formation of vacuoles; above 100 °C: beginning of vaporization and tissue carbonization.

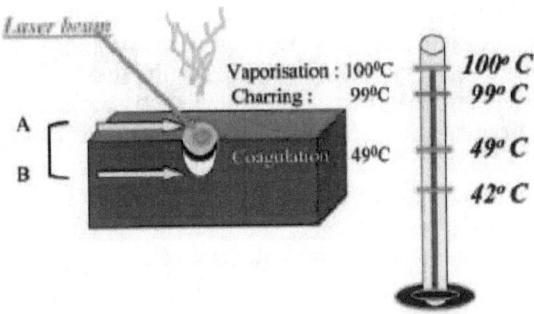

Fig. 2.1 Tissue effects after laser beam interaction

Thermal response of the tissue after laser irradiation is dependent of a number of variables, but the optical tissue properties play a dominant role. Thermal effect is based on the absorption of radiation and laser energy conversion into heat. Heat energy is deposited in the tissue by the absorption of light and its subsequent conversion to heat via vibrational relaxation. This causes a rise in temperature of the tissue. Then the heat will diffuse through the tissue causing a rise in temperature in the surrounding tissue. The damage done to the tissue depends on the temperature that is reached, and the duration at which it is held at that temperature. First, we look at how we can calculate the temperature rise in the tissue and how the heat diffuses to the surrounding tissue. Then we look at the effect of the raised temperature on different types of tissue - the damage done by the heating. Depending on the temperature rich in the different zone of the target tissue can appear histological modification: denaturation, coagulation, conformational changes, cellular death.

When a tissue is exposed to the laser irradiation the tissue temperature increases due to the absorption process. Heat effects depend on the type of tissue and the temperature achieved inside the tissue [1-7]. Temperature is the governing parameter of all thermal laser-tissue interactions. For the purpose of predicting the thermal response, a model for the temperature distribution inside the tissue must be derived. Different types of lasers react differently with tissue and the interaction depends on: the wavelength of the laser, power density and exposure time, optical and thermal property of the tissue being irradiated, laser beam size on the tissue and if the exposure is CW or pulsed wave radiation. The following factors will affect the absorption of laser light by a target tissue:[8] laser wavelength, tissue (composition), tissue thickness, surface wetness, incident angle of beam, exposure time, contact vs non-contact modes, laser incident power (J/s), laser power density (W/cm^2): for any chosen level of incident power, the smaller the beam diameter, the greater concentration of heat effects, beam movement: relative to tissue site; rapid laser beam movement will reduce heat build-up and aid thermal relaxation, endogenous coolant: blood flow, exogenous coolant: water, air, pre-cooling of tissue.[6,7]

The spatial distribution of soft tissue damaged depends on the temperature distribution. The equation who describes the thermal diffusion, the temperature spread in the tissue is called the heat diffusion equation [5-8].

Mathematical modeling of heat transfer in biological tissues can be done by studying the bioheat transfer equation. Solving analytical and/or numerical equation of temperature distribution $T = T(r,t)$ at each point determined by the position of r vector at t time. When the tissue temperature in tissue is closer to the vaporization threshold, photothermal effects appear under the influence of:
- the energy required to change the water phase;
- dehydration of the tissue;
- formation of the steam vacuole in the tissue;
- mechanical effects due to rapid expansion of water vapors.

2.2 Modeling heat transfer in tissue

The computer software, Comsol Multiphysics, makes it possible to numerically solve partial differential equations [5-12]. The numerical solution relies on the Finite Element Method (FEM), in which the geometry studied is divided into a finite element mesh. Thus, instead of trying to solve a highly non-linear problem on the entire geometry, an approximate solution is sought in each element. If this element is considerably small the physical problem is assumed to vary linearly.

Geometry

One of the main motivations in applying the FEM for solving light propagation problems is the ability to use arbitrary geometries. There exist numerous of different ways to import geometries into Multiphysics, although in this exercise we will use simple shapes that are drawn by hand. After the geometry has been defined, a mesh is created. In Multiphysics this is automated to a great extent but some settings are important to change. The accuracy of the FEM solution depends on the mesh quality. Hence it is of utmost importance that the mesh is fine when the gradient of the fluence rate is large. In practice this means that the mesh should have a higher resolution around sources and if possible also around detectors.

In the context of tissue heat treatment the four principal modes of heat transfer are conduction, convection, radiation and evaporation. This section briefly describes the different modes. Conduction is an intrinsic property of the medium, whereas evaporation acts as a boundary condition at tissue-air boundaries. Radiation and convection are present to varying degrees both inside the bulk tissue and at surfaces.

Conduction

Thermal conduction is transfer of energy from the more energetic particles in a substrate to the adjacent less energetic ones as a result of interactions between the particles. This mode of tissue heat transfer is described by Fourier's law, which states that the heat flow at a given point in the tissue is proportional to the gradient of the temperature. An energy balance for a volume element gives the following relation for transient conduction

$$\rho c \frac{\partial T}{\partial t} = \nabla(\lambda \nabla T) \qquad (2.1)$$

where, ρ (kg m^{-3}) is the density, c (J kg^{-1}K^{-1}) is the specific heat capacity, λ (W m^{-1}K^{-1}) is the thermal conductivity, and T (K) is the tissue temperature. The ratio $\lambda/(\rho c)$ is called the thermal diffusivity, α (m^{2}s^{-1}), which describes the dynamic behavior of the thermal process.

The thermal properties depend on the type of tissue and temperature. Since most tissues can be considered as being composed of a combination of water, proteins and fat, the magnitude of the conductivity can be estimated with knowledge of the proportions of these components. Table 1 shows selected data on the thermo-physical properties of tissue measured at room or body temperature. It can be observed that of the different components constituting tissue, water has the highest values of the thermal conductivity and diffusivity. Therefore, tissues with high water content are the best thermal conductors and respond the fastest to a thermal disturbance.

Table 2.1 Thermophysical properties of human tissue and water

Material	Conductivity (Wm^{-1}K^{-1})	Density (kg m^{-3})x10^{-3}	Specific heat (kJkg^{-1}K^{-1})	Diffusivity (m^{2}s^{-1})x10^7
Muscle	0.38-0.54	1.01-1.05	3.6-3.8	0.90-1.5
Fat	0.19-0.20	0.85-0.94	2.2-2.4	0.96
Kidney	0.54	1.05	3.9	1.3
Heart	0.59	1.06	3.7	1.4
Liver	0.57	1.05	3.6	1.5
Brain	0.16-0.57	1.04-1.05	3.6-3.7	0.44-1.4
Water@37°C	0.63	0.99	4.2	1.5

Convection

Convection is the term applied to heat transfer between a surface and a fluid, e.g., air or blood, moving over the surface. Convection problems involve heat transfer between the surface of the body and the surrounding air and also between blood and the vessel wall. The heat transfer between a conducting solid, which in this context is tissue, and a convecting fluid is usually described using Newton's law of cooling which states that the heat flow q (W m^{-2}) normal to the surface is proportional to the temperature difference between the surface T (K) of the tissue and the bulk temperature of the fluid T_∞: $q = h(T - T_\infty)$

where, the proportionality constant, h (W m^{-2}K^{-1}), is called the local heat convection coefficient.

At the tissue surface there is no fluid motion and energy transfer occurs only by conduction. The magnitude of the heat convection coefficient therefore depends on the temperature gradient at the surface but also on the conditions in the boundary layer. Equations can be set up to account for the physical processes in the boundary layer but the solutions readily become very complex. Here it suffices just to give examples of the convection coefficient in a few cases:

Table 2.2 Values of the convection coefficient for a few cases

Mode	h(W/m²/K)
Free convection (tissue-air)	5-25
Forced convection (tissue – air, i.e. a windy day)	50-20 000
Convection with phase change (boilong)	2500 – 100 000

Radiation

Every body emits electromagnetic radiation proportional to the fourth power of the absolute temperature. The maximum heat flow, q (W m^{-2}) emitted from a black body is expressed by:

$$q = \varepsilon\sigma(T_s^4 - T_\infty^4) \qquad (2.2)$$

where, σ (Wm^{-2}K^{-4}) is the Stefan-Boltzmann constant, T_s and T_∞ (K) are the tissue surface and environmental temperature, respectively and ε is the emissivity. Since tissue is not a perfect black body the emissivity is less than one. For human skin, the value of the emissivity is in the range 0.98-0.99. When modeling internal biological tissue, intrinsic radiative heat transfer processes constitute a negligible contribution and the process of radiation is often considered only a boundary effect.

Evaporation

The heat loss from moist tissue in contact with air will be dominated by heat loss associated with evaporation of water. The driving force for the loss of tissue moisture by diffusion is the difference in water concentration between the tissue and the surrounding air. For ideal gas mixtures, such as air and water vapors, the concentration of a component is proportional to the partial pressure. The water at the tissue surface is in thermodynamic equilibrium with the water vapors in air, i.e., at the surface the pressure of the water vapors is equal to the saturated water vapors pressure at the surface temperature. Therefore, water will migrate from the tissue surface to the air if the saturated vapors pressure at the surface is greater than the partial pressure of vapors in the surrounding air. Evaporation acts as a loss term in the bio-heat equation at tissue surfaces.

Bio-heat equation

The bioheat equation for heat transfer in tissue requires information regard to the tissue geometry (shape, structure, etc) and laser beam (beam type). The heat transport in living tissue after interaction with laser radiation is influenced by the following factors:
- tissue thermophysical properties (heat capacity, thermal conductivity, etc.);

- geometry of the irradiated tissue;
- production of heat due to the absorption of laser radiation;
- production of heat due to metabolic processes;
- heat transport by the blood convection (blood perfusion;
- thermoregulatory mechanisms in the tissue.

Each of the factors described above re included in the equation of heat transfer in tissue within a separately so that it can be done when the situation requires, simplify the equations by neglecting certain terms.

Heat transfer equation is in fact a law of energy conservation. Inside tissue, the heat transfer is described by the bio-heat equation taking into account conduction, convection by blood and possible heat sources.

In the thermal conduction process the energy is transferred from the more energetic particles in a substrate to the adjacent less energetic ones as a result of interactions between the particles. This mode of tissue heat transfer is described by Fourier's law, which states that the heat flow at a given point in the tissue is proportional to the temperature gradient [5-9]:

$$\rho C \frac{\partial T}{\partial t} = \nabla(k \nabla T) \qquad (2.3)$$

where, ρ (kg m^{-3}) is the density, C (J kg^{-1}K^{-1}) is the specific heat capacity, k (W m^{-1}K^{-1}) is the thermal conductivity, and T (K) is the tissue temperature.

The thermal properties depend on the type of tissue and temperature and, the tissues can be considered as being composed of a combination of water, proteins and fat, the magnitude of the conductivity can be estimated with knowledge of the proportions of these components. Convection is the term applied to heat transfer between a surface and a fluid, e.g., air or blood, moving over the surface, and radiation, which has to do with the infrared energy that any body at nonzero temperature emits naturally.

The mathematical model used to investigate the laser-tissue interaction is based on the Pennes bio-heat equation [6]:

$$\rho C \frac{\partial T}{\partial t} + \nabla(-k \nabla T) = \rho_b C_b \omega_b (T_b - T) + Q_{met} + Q_{laser} \qquad (2.4)$$

where, ρ, C and k are the density, specific heat and thermal conductivity of the tissue, Q_{met} is the metabolic heat generation per unit volume, ρ_b, C_b are the density and the specific heat of blood, ω_b is the blood perfusion rate, T_b the temperature of arterial blood, T is the local tissue temperature and Q_{laser} (W/m^3) is the heat source.

$$Q_{laser} = \mu_a q \qquad (2.5)$$

where μ_a (m^{-1}) is the absorption coefficient and q (W/m^2) is the power density.

The two main mechanisms for heat flow inside a tissue are: through conduction, meaning that the gradient in temperature within the tissue itself drives the flow, and through convection of thermal energy by the blood perfusion.

Because of desirable features such as high precision, less blood loss and minimal invasiveness, carbon dioxide (CO_2) lasers are widely used as cutting tools [1, 2] in many medical fields. In treatments, that use laser radiation it is important to have a control of temperature, high temperature could lead to thermal damage in the surrounding tissue [1,2]. The deposition of laser energy in the tissue is not only a function of laser parameters such as wavelength, power density, exposure time, spot size, and repetition rate [1], it also depends on opto - thermal tissue parameters. To choose the treatment parameters properly and to predict the outcome of the photothermal effect, a reliable tissue model and simulation method is needed. Temperature is the governing parameter of all thermal laser– tissue interactions [2]. Depending on the tissue temperature, different effects are observed. For example, the denaturation of the enzymes and the looseness of the membranes occurs at 40-45^0 C; coagulation, necrosis and protein denaturation occurs around 60^0 C; drying out occurs at 100^0 C; carbonization occurs at 150^0 C, and, finally, pyrolysis and vaporization occurs at above 300^0 C. Among those temperatures, 45^0 C is a characteristic temperature used frequently in photodynamic therapy. Increasing the tissue temperature to 100^0 C boiling of tissue water is induced. For continued heating of dehydrated tissue to about 400^0 C an initiated burning with the production of char tissue and smoke it is observed. Therefore, for the purpose of predicting the thermal response, a model for the temperature distribution inside the tissue must be derived.

In this paper, the soft tissue model is used to predict the temperature distribution during CO_2 laser treatment. Finite element method (FEM) is applied for modeling of numerical simulation to predict the dynamic of temperature in human tissue using Pennes bioheat equation. The objective of this research is modeling the laser-tissue interaction, to optimize the effective parameters in order to understand optimal laser dosage and to prevent the tissue damage. Numerical simulations are compared with the OCT (Optical Coherent Tomography) images of the pig vocal cords irradiated zone by a CO_2 surgical system.

2.3 Mathematical model

To generate the finite element model we used the software Comsol Multiphysics [4]. This software provides all the elements needed to build the model, solve the problem and post-process the results. Solutions in simple cases (simple geometry, independent of temperature or uniform tissue properties) guides us to understand how the heat behave. The numerical solution relies on the Finite Element Method, in which the geometry studied is divided into a finite element mesh with computer software that makes it possible to numerically solve partial differential equations. Thus, instead of tying to solve a highly non-linear problem on the entire geometry, an approximate solution is sought in each element. If this element is considerably small the physical problem is assumed to vary linearly. Numerical simulation helps us in obtaining the desired result according to the laser parameters and pathology [5,6]. The heating of tissue during laser ablation is modeled by the bioheat equation (Eq. 2.6).

The light absorption in the tissue was modeled as an exponential decay using Beer-Lamberts law:

$$I = I_0 \exp(-\mu_a z) \quad (2.6)$$

where, $I_0 = \dfrac{P}{A}$ is the intensity of laser, A (m²) is the area of the tissue and P (W) is the output power of the laser.

Tabel 2.3 The optical and thermal properties for soft tissue used in numerical simulation

Parameters	Value
Specific heat of tissue - C	3600 J/kg*K
Density of tissue - ρ	1050 kg/m³
Thermal conductivity tissue - k	0.5 W/m*K
Blood perfusion rate tissue - ω_b	0.0005ml/s/ml
Arterial blood temperature - T_b	37 °C (310.15 K)
Specific heat of blood - C_b	4200 J/kg*K
Density of blood - ρ	1000 kg/m³
Heat coefficient transfer - h	10 W/m²*K
Surrounding air temperature	23 °C (296.15 K)

To accomplish our model we have made different stapes. The first step was to choose the geometry of the model, tissue sample being a cylinder. Then we have defined the physical properties and the boundary conditions as well as Q_{laser}. The external heat source is due to the absorption in the stained tissue, i.e. $Q_{laser} = 0$ everywhere except in the cylindrical subdomain representing the laser spot. The laser spot was considered as a cylindrical sub-domain, where heat source was confined. Boundary conditions depend on the particular problem and, the most usual case is a biological tissue irradiated from the outside and the laser light propagated in a direction perpendicular to the tissue. In this case continuity is commonly assumed, which means that the gradient of the temperature with respect to the coordinate affected by the boundary is zero ($\partial T/\partial x = 0$; $\partial T/\partial y = 0$; $\partial T/\partial z = 0$). Initial and boundary conditions can be set by using the steady-state temperature of biological tissue [8,9], $t = 0$, $T = T_b$, i.e., the initial temperature of the tissue is equal to the blood temperature. The initial temperature at $t = 0$, was set at 310.15 K (37 °C). The heat generation by metabolism and heat transfer by blood perfusion are constant in biological tissue [8,9]. The laser beam profile used in our experiments and numerical analysis was a Gaussian profile, which means that the concentration of energy focuses on its center section. The spot diameter $d = 800$ µm of the laser is the same used in the experimental analysis; the incident beam is perpendicular to the tissue surface and the absorption coefficient used is $\mu_a = 767$ cm^{-1} [10,11], which correspond to the absorption coefficient of 10.6 µm in soft tissue. Moreover, the following assumptions were adopted: thermal radiation emission at the tissue/air interface and reflection of laser light from the external surface of the cornea were disregarded; the thermo-optical parameters were considered to be constant during the process. The bioheat equation is solved for a single layer model of the soft tissue to predict the temperature distribution using CO_2 laser light source radiation burns at as steady state. The output is the real time temperature of the tissue. Plots of temperature changes with time can also be derived [10,11].

Steps in Comsol numerical simulation:

1. Mechanism choice.

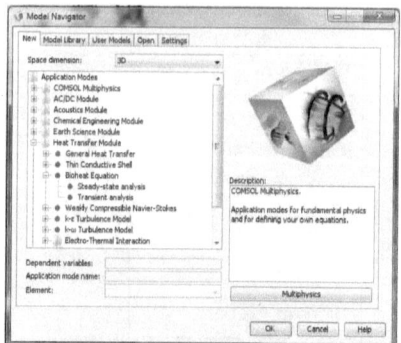

2. Geometry

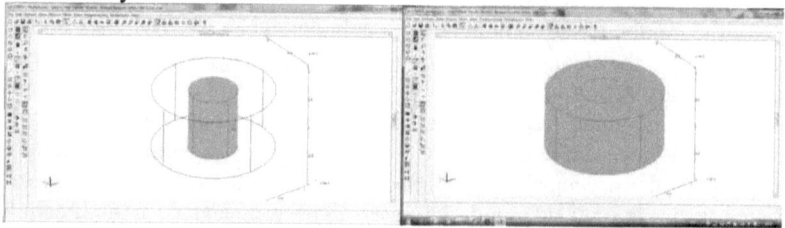

3. Tissue and blood parameters

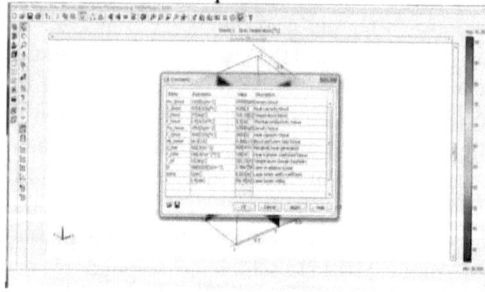

4. Introducing the formula for calculating the heat generated by laser radiation

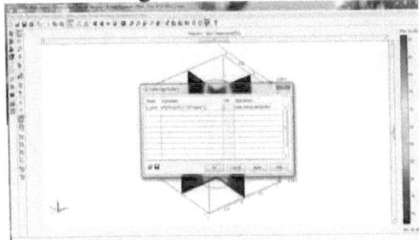

5. Subdomain setting

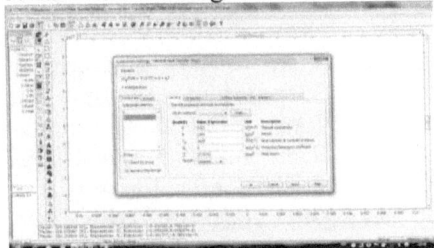

6. Setting the boundary condition

7. Initialize mesh

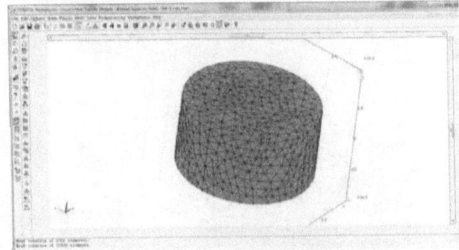

8. Solving the problem

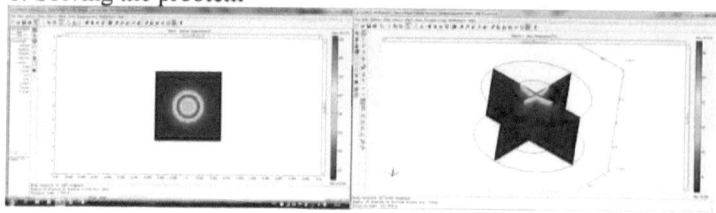

2.4 Results and discussions

Temperature dynamics during laser tissue ablation was studied experimentally by irradiation of larynx tissue with a CO_2 laser surgical system at different powers and different exposure times. Experimental analysis of the crater ablation is presented in Fig. 2.2 by the OCT images, which show the thermal damage in biological tissue after laser irradiation at different powers (6W, 14 W, 20 W and 24 W) and an exposure time of 100 ms. It can be observed that the ablation volume and the thermal damage in the surrounding tissue increase with the power of the laser beam. The irradiation area is indicated by a bleak ring on the figures. The ablation crater has a conic shape, with an input aperture of about 800 µm diameter and a depth of ~ 0.64 mm at 14 W, 0.86 mm at 20 W and 0.92 W at 24 W, which is in accordance with the focal length of the laser beam focusing system ($f = 125$ mm) and with the laser beam divergence. At the lower power of the laser beam i.e., the 6 W, the irradiation effects are observed only as a necrosis zone there is no ablation crater in the tissue (see Fig. 2.2a.).

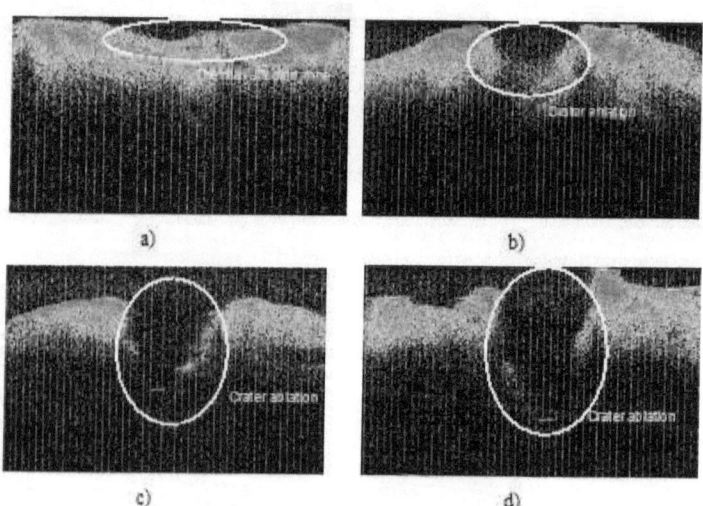

Fig. 2.2 OCT images of the crater ablation in porcine vocal cord using a CO_2 at $t_{exp} = 100$ ms at laser power: (a) P= 6 W, (b) P = 14 W, (c) P = 20 W, (d) P = 24 W

During laser treatment it is very important to study the temperature rise in the laser irradiated wound area, and to evaluate the temperature in the surrounding tissues does not exceed the threshold of irreversible thermal damage.

Numerical analysis of 3D temperature distribution in CO_2 irradiated soft tissue was studied [12-18]. For the numerical simulations were used the laser parameters used *in vitro* to ablate the tissue, the wavelength of 10.6 μm (CO_2 laser wavelength), the corresponding absorption coefficient in the tissue, laser powers of 6 W, 14 W, 20 W and 24 W, an exposure of tissue to the laser beam of 100 ms and a laser beam with diameter of 800 μm. The tissue parameters used are presented in Table 2.3. The CO_2 laser has a small penetration depth in the tissue of 15 – 20 μm [1], and because of the small depth of penetration it was assumed that the irradiated tissue layer is homogeneous.

The computational models are simulated and presented in Fig. 2.3-2.6. Fig. 2.3-2.6 shows the three-dimensional temperature distribution in irradiated tissue at 6 W, 14 W 20 W and 24 W at an irradiation time of 100 ms, and the temporal evolution of temperature with depth. At laser power of 6 W and exposure time of 100 ms the maximum temperature is 67^0 C (see Fig. 2.3). At this temperature there is no laser ablation in the tissue, is presented coagulation, necrosis and protein denaturation, the tissue is thermal affected, and this can be seen also in the OCT image (see Fig. 2.2(a)). By increasing the laser power, the increased temperature in the tissue, at 107^0 C tissues drying and at 137^0 C and 160^0 C tissues carbonization (see Fig. 2.5 and 2.6). From the numerical simulation at 14 W, 20 W and 24 W can be seen that the temperature exceeds 100^0 C with the ablation crater that is present even in the OCT images (see Fig. 2.2) with the crater ablation irradiation results.

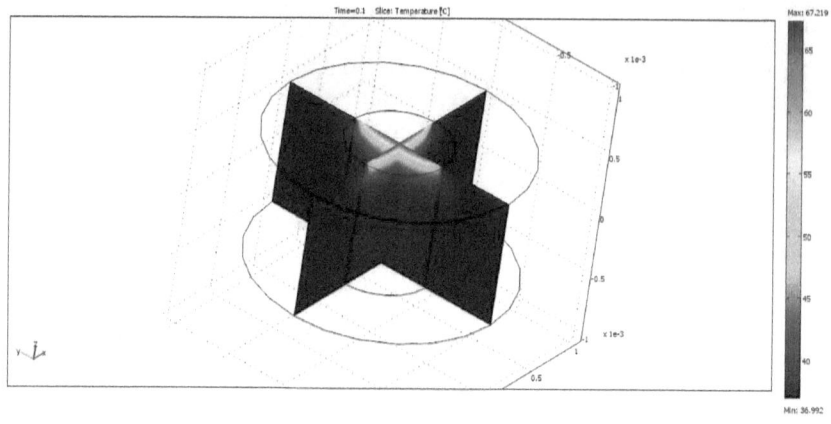

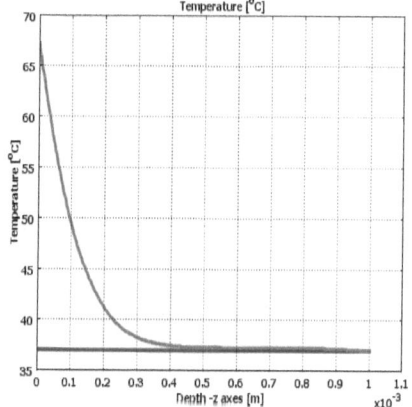

Fig. 2.3 Temperature distribution in tissue irradiated at P = 6 W and t_{exp} = 100 ms: 3D temperature distribution; temporal evolution of temperature with depth, $T(x = 0, y = 0, z, t = 100$ ms$)$

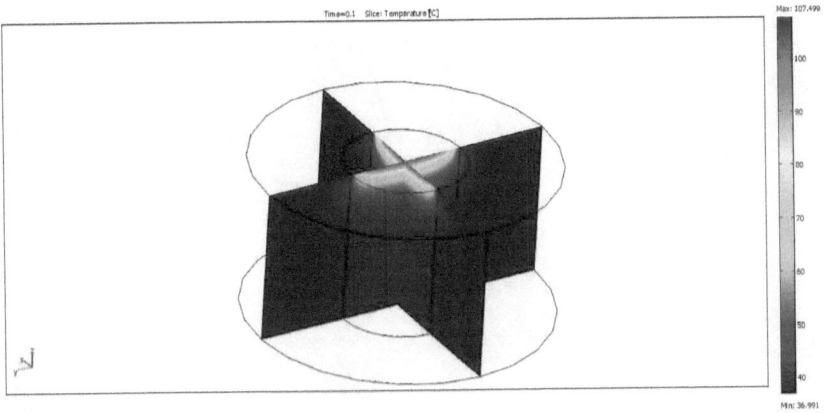

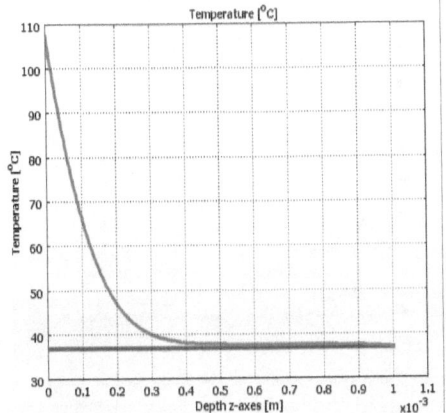

Fig. 2.4 Temperature distribution in tissue irradiated at P = 14 W and t_{exp} = 100 ms: 3D temperature distribution; temporal evolution of temperature with depth, $T(x = 0, y = 0, z, t_{exp} = 100$ ms$)$.

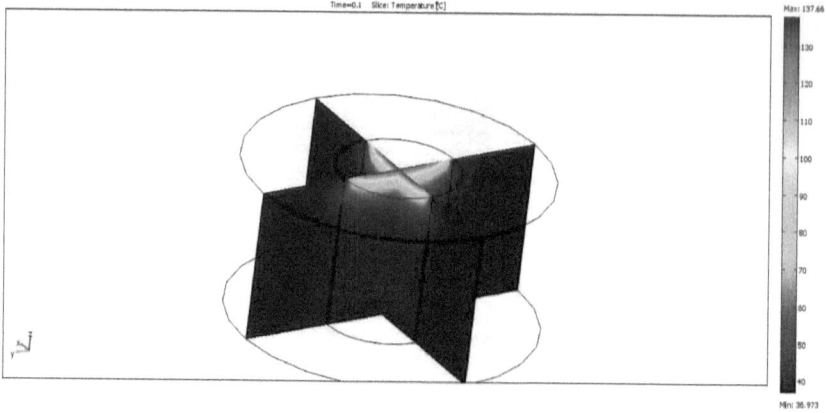

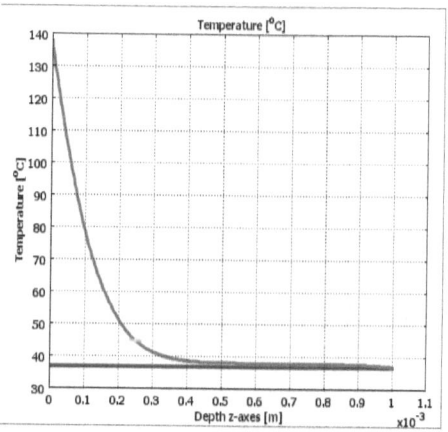

Fig. 2.5 Temperature distribution in tissue irradiated at P = 20 W and t_{exp} = 100 ms: 3D temperature distribution; temporal evolution of temperature with depth, $T(x = 0, y = 0, z, t_{exp} = 100$ ms).

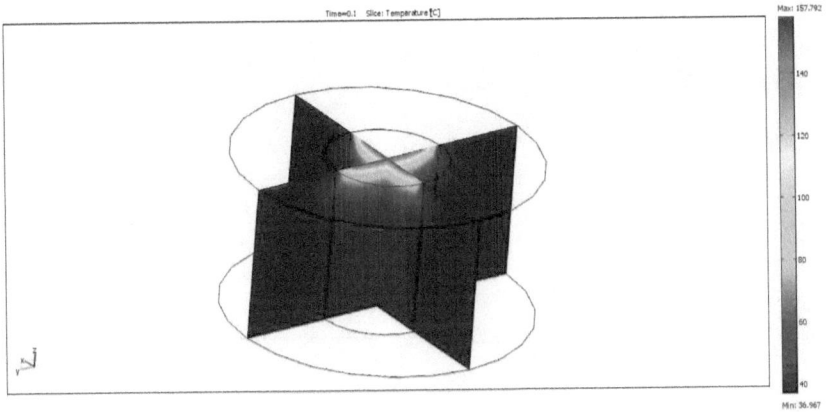

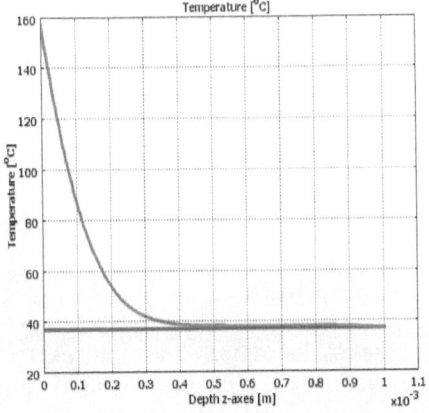

Fig.2.6 Temperature distribution in tissue irradiated at P = 24 W and t_{exp} = 100 ms: 3D temperature distribution; temporal evolution of temperature with depth, $T(x = 0, y = 0, z, t_{exp} = 100$ ms).

The temperature evolution in time in irradiated tissue at 6 W, 14 W, 20 W and 24 W is presented in Fig. 2.7. It can be observed that the temperature tends to a constant steady-state value for the irradiation time longer than 1 s [13]. Therefore, the exposure time is an important parameter in the medical laser applications, especially for very short exposure time. As one can observed, the time needed to reach the constant steady-state value temperature is independent of the optical source power. However, the value of the constant temperature depends on the laser power. Due to this increase in the tissue temperature; one must optimize with a high accuracy the two laser parameters: the exposure time and the power of the laser.

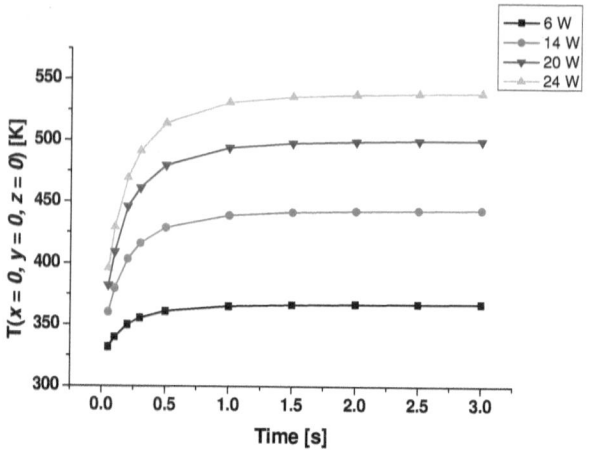

Fig. 2.7 Temporal evolution of temperature at laser impact point T($x = 0$, $y = 0$, $z = 0$), different laser powers: 6 W, 14 W, 20 W and 24 W and an irradiation time of 100 ms

2.5 Conclusions

In this work, a 3D numerical approach for the bio-heat equation has been presented and the numerical procedure has been analyzed. The model was applied to the irradiated soft tissue and the results compared with the experimental OCT dates. We simulate the temperature evolution in biological tissue using the wavelength of 10.6 μm which is strongly absorbed by the water and has a small penetration depth in biological tissue of ~ 15 – 20 μm. In this case, the tissue parameter homogeneity remains constant over time (C, k, ρ). The numerical simulations are in good agreement with the OCT images obtained after *in vitro* irradiation of pig vocal cords.
The results obtained from the study reveal the important parameters, which significantly affect the thermal response of soft tissue, as exposure duration, power, wavelength of the beam, as well as the area and type of the tissue. The exposure time is an important parameter in the medical laser application and an increase of the temperature in the tissue after a long exposure and propagation of the heat in the surrounding tissue leads to thermal damage.
The numerical method of temperature distribution modeling can also be used for extensive parametric studies in order to characterize the stability of various treatment parameters and would allow obtaining a faster and better simulation of laser treatment of biological tissues. Solving the bioheat equation numerically, allows obtaining useful information for analysis of different laser surgical techniques. The present model could serve as a good tool to predict temperature distribution in tissues under laser irradiation.

References

1. *R.H. Ossoff, L. Reinisch,* Chapter 12: Laser Surgery: Basic Principles and Safety Considerations, C.V. Mosby Company, St. Louis (1986).
2. *M.H. Niemz,* Laser-tissue interactions, Fundamentals and Applications, Third Enlarged Edition, Springer (2007).
3. *R. Steiner,* Laser and IPL Technology in Dermatology and Aesthetic Medicine, Chapter: Laser-tissue interactions, Springer-Verlag Berlin Heidelberg (2011).
4. *C-S. Orr, R.C. Eberhart,* Optical thermal response of laser-irradiated tissue, Chapter 11 – Overwiew of bioheat transport, p.:367-384, Plenum Pres, New York (1995).
5. Light propagation modelling using Comsol Multiphysics Medical Optics Course, Atomic physics Lund University, Version 1.0 Biophotonics Group, Lund University: http://www.atomic.physics.lu.se/fileadmin/atomfysik/Biophotonics/Education/MultiphysicsExercise.pdf
6. *J.P. Holman, Heat Transfer,* MC Graw Hill (1981).
7. *J-L. Dillenseger, S. Esneault,* Fast FFT-based bioheat transfer equation computation, Computers in Biology and Medicine, 40:119-123 (2010).
8. *A. Sakurai, I. Nitta, S. Maruyama, J. Okajima,* K. Matsubara, Coupled photon and heat transfer simulation inside biological tissue for laser therapy, Journal of Thermal Science and Technology, 4(2): 314-323 (2009).
9. http://www-atom.fysik.lth.se/MedOpt/FEM/MultiPhysicsExercise.zip/
10. *M. Petrus, E.M. Carstea, D. C. A.* Dutu, *D. C. Dumitras,* Optical coherence tomography monitoring of laser ablation processes in ENT tissues, Journal of Optoelectronics and Advanced Materials, 13 (7): 776 – 780 (2011).
11. *M. Petrus, D.C. Dumitras,* Numerical simulation and in vitro experimental temperature distribution analysis in irradiated tissue, U.P.B Sci. Bull., Series A , 74 (4): (2012).
12. *F. Rossi, R. Pini, L. Menabuoni,* 3D Simulation and Experimental Comparison of Temperature Dynamics in Laser Welded Cornea, Proceedings of the COMSOL Users
13. *B. Choi, A.J. Welch,* Analysis of thermal relaxation during laser irradiation of tissue, Lasers in Surgery and Medicine, 29: 351-359 (2001).
14. *J.T. Walsh, J.P. Cummings,* Effect of the dynamic optical properties of water on midinfrared laser ablation. Lasers Surg Med, 15: 295-305 (1994).
15. *B. Choi, J.K. Barton, E.K. Chan, A.J. Welch,* Imaging of the irradiation of skin with a clinical CO_2 laser system: implications for laser skin resurfacing. Lasers Surg Med, 23:185-0193 (1998).
16. *E.D. Jansen, R.K. Chundru, S.A. Samanani, T.A. Tibbetts, A.J. Welch,* Pulsed infrared laser irradiation of biological tissue: effect of pulse duration and repetition rate. Proc SPIE, 1882:322-326 (1993).
17. *K. Schomacker, J.T. Walsh, T. Flotte, T. Deutsch.* Thermal damage produced by high-irradiance continuous wave CO_2 laser cutting of tissue. Lasers Surg Med, 10:74-84 (1990).

18. *A.D. Zweig, B. Meierhofer, O.M. Muller, C. Mischler, V. Romano, M. Frenz, H.P. Weber*. Lateral thermal damage along pulsed laser incisions. Lasers Surg Med, 10:262-274 (1990).

Chapter 3
Spectroscopic analysis of surgical smoke produced *in vitro* by laser vaporization of animal tissues in a closed gaseous environment

3.1. Introduction

The use of thermal instruments for surgical applications has grown significantly over the past three decades. Unfortunately, when any type of energy-based surgical instrument such as laser, high-frequency electric knives, and ultrasonic/harmonic scalpel is applied to human tissue, an unwanted by-product is produced, commonly known as surgical smoke. Surgical smoke is created when the laser energy is delivered to a tissue, owing to the heating of the tissue components. The heat vaporizes the intracellular fluid, which increases the pressure inside the cell and causes the cell membrane to burst. When this happens, a plume of smoke containing mostly water vapor is released into the atmosphere of the operating room. At the same time, the intense heat creates chars in the protein and other organic matter within the cell, and causes thermal necrosis in adjacent cells. The charring of cells releases other harmful contaminants, such as carbonized cell fragments and gaseous hydrocarbons in the form of aerosols, vapors and gases [1-17]. The gases and vapors produced during laser tissue interaction consist of [1-17]:
- combustion degradation gases, such as carbon dioxide (CO_2), carbon monoxide, hydrogen sulfide, and ammonia;
- volatile organic compounds, such as toluene, styrene, methylpyrazine, benzaldehyde, indol, skatol, phenol and benzyl cyanide;
- volatile organic compounds, such as formaldehyde, benzene, ethanol, carbon disulfide, well known as carcinogen.

Aerosols, vapors and gases in the surgical smoke produced during laser vaporization of the tissue may constitute a threat to personnel and the patient health and can be a potential risk. Besides the health risks due to the volatile organic compounds, particulate matter and tissue fragments ejected into the atmosphere can be inhaled and represent an additional risk. The noxious odor of surgical smoke is an indication of the content of the smoke. The smell is a conglomeration of chemical by-products from the burning of proteins and lipids when using laser or electrosurgical instruments.

The aerosols form the surgical smoke has been shown to be a viable transport mechanism for viruses, blood and cells [1-17]. Although rare, there have been reports of diseases transmitted from the patient to the surgeon via surgical smoke. From the health and safety perspective, the chemical, biological and particulate compositions of surgical smoke are therefore of great interest. The risk of surgical smoke is due to the odor, size of particles and gas concentration. The great threat is represented by the toxins that cause the odor in surgical smoke. The smell is a combination of chemicals from the combustion of proteins and lipids when laser is used. Studies have shown that these chemicals cause headaches, eye, nose and throat irritation, as well as potential long-term effects. Some of these toxic gases have already been shown to be carcinogenic, such as benzene, which has been documented to be a trigger for leukemia. Benzene was found in surgical smoke in concentration well above the permissible limit. Beside benzene, during laser tissue ablation, the released chemicals include formaldehyde, acrolein, carbon monoxide and hydrogen cyanide. High concentrations of carbon dioxide are also potentially dangerous. Analysis of surgical smoke is important in evaluation of risk to at which the medical staff and the patients are exposed. The CO_2 laser is used by various medical specialties to vaporize, ablate or cut tissue. The CO_2 laser produces light with a wavelength of 10.60 μm, where the laser energy is highly absorbed by water and by all biological tissues with rich water content. When the absorbed laser light heats the tissue to approximately 100°C, vaporization of intracellular water occurs. This produces vacuole formation, crater, and tissue shrinkage. As the CO_2 laser beam comes into contact with the tissue, the temperature rises and causes boiling the intra- and extracellular water. The cell expands due to steam formation, and explodes. The contents are released as steam and smoke. Surgical smoke is made of chemicals, blood, tissue particles, viruses, bacteria and consists mostly of water vapor and carbon dioxide. The size of particles formed varies between over 200 μm and less than 10 nm. The mean diameter of particles depends in particular on how intensely the energy applied acts on tissues. The mean diameter of particle from surgical smoke after laser ablation is ~ 0.3 μm. The effect of surgical smoke is estimated from its composition and the noxious nature of its components.

In this paper we present a study of chemical composition (six gases) of surgical smoke produced *in vitro* by irradiation of fresh animal tissue using a CO_2 laser in nitrogen atmosphere. Acetonitrile, acrolein, ammonia, benzene, ethylene, and toluene were the gases that we determined using our CO_2 photoacoustic spectroscopy system. The influence of laser power, exposure time and the type of tissue on gas concentrations was investigated.

3.2 Experimental section: Laser Photoacoustic Spectroscopy

Laser photoacoustic spectroscopy (LPAS) has emerged over the last decade as a very powerful investigation technique, capable of measuring trace gas concentrations at ppmV (parts per million by volume), or even sub-ppbV (parts per billion by volume) level [18-29]. LPAS provide several unique advantages, notably the multicomponent capability, high sensitivity and selectivity, wide dynamic range, immunity to electromagnetic interferences, convenient real time data analysis, operational simplicity, relative portability, relatively low cost per unit, easy calibration, and generally no need for sample preparation. LPAS is primarily a calorimetric technique and, as such, differs completely from other previous techniques, as the absorbed energy can be determined directly, instead of via measurement of the intensity of the transmitted or backscattered radiation. In conjunction with tunable lasers, *in situ* monitoring of many substances occurring at ppbV or even pptV (parts per trillion by volume) concentrations is a routine task today. Photoacoustic (PA) detection provides not only high sensitivity but also the necessary selectivity for analyzing multicomponent mixtures by the use of line-tunable IR lasers

CO_2 LPAS offers a sensitive technique for detection and monitoring of trace gases at low concentrations. The CO_2 laser is of special interest, as it ensures high output power in a wavelength region (9-11 μm) where more than 250 molecular gases/vapors of environmental concern for atmospheric, industrial, medical, military, and scientific spheres exhibit strong absorption bands. This laser, however, can be only stepwise tuned when operated in cw. Nevertheless, it is an ideal source to push the sensitivity of PA gas detection into the concentration range of ppbV or even lower. Instruments based on LPAS have nearly attained the theoretical noise equivalent absorption detectivity of 10^{-10} cm^{-1} in controlled laboratory conditions.

PA spectroscopy is an indirect technique in that an effect of absorption is measured rather than absorption itself. Hence the name of photoacoustic: light absorption is detected through its accompanying acoustic effect. The advantage of photoacoustics is that the absorption of light is measured on a zero background; this is in contrast with direct absorption techniques, where a decrease of the source light intensity has to be observed. The spectral dependence of absorption makes it possible to determine the nature of the trace components. The PA method is primarily a calorimetric technique, which measures the precise number of absorbent molecules by simply measuring the amplitude of an acoustic signal. In LPAS the nonradiative relaxation which generates heat is of primary importance. In the infrared spectral region, nonradiative relaxation is much faster than radiative decay. The sample gas is in a confined (resonant or nonresonant) chamber, where modulated (e.g., chopped) radiation enters via an IR-transparent window and is locally absorbed by IR-active molecular species. The temperature of the gas thereby increases, leading to a periodic expansion and contraction of the gas volume synchronous with the modulation frequency of the radiation. This generates a pressure wave that can be acoustically detected by a suitable sensor, e.g., by a microphone. The advantages of the PA method are high sensitivity and small sample volume; besides, the acoustic measurement makes optical detection unnecessary. The main drawback is caused by the sensitivity to acoustic noise, because the measurements are based on an acoustic signal.

The favorable properties of LPAS are essentially determined by the characteristics of the laser. The kind and number of detectable substances is related to the spectral overlapping of the laser emission with the absorption bands of the trace gas molecules. Thus, the accessible wavelength range, tunability, and spectral resolution of the laser are of prime importance. With respect to minimum detectable concentrations (LPAS sensitivity), a laser with high output power P_L is a benefit, because the PA signal is proportional to P_L. The broad dynamic range is an inherent feature of LPAS and therefore is not affected by the choice of the radiation source. In contrast to remote-sensing methods, LPAS is a detection technique applied locally to samples enclosed in a PA cell. In order to still obtain some spatial resolution, either the samples have to be transported to the system, or the system has to be portable. The temporal resolution of LPAS is determined by the time needed for laser tuning and the gas exchange within the cell. Thus, a small volume PA cell and a fast tunable laser are a plus.

The availability of suitable laser sources plays a key role, as they control the sensitivity (laser power), selectivity (tuning range), and practicability (ease of use, size, cost, and reliability) that can be achieved with the photoacoustic technique. The CO_2 laser perfectly fits the bill for a trace gas monitoring system based on LPAS. This IR laser combines simple operation and high output powers. The frequency spacing between two adjacent CO_2-laser transitions range from 1 to 2 cm^{-1}. By contrast, the typical width of a molecular absorption line is approximately 0.05 to 0.1 cm^{-1} for atmospheric conditions. Since this is not a continuously tunable source, coincidences between laser transitions and trace gas absorption lines are mandatory. Fortunately, this does not hamper its applicability to trace gas detection, as numerous gases exhibit characteristic absorption bands within the wavelength range of the CO_2 laser which extends from 9 to 12 μm when different CO_2 isotopes are used. The CO_2 laser spectral output occurs in the wavelength region where a large number of compounds (including many industrial substances whose adverse health effects are a growing concern) possess strong characteristic absorption features and where absorptive interferences from water vapors, carbon dioxide, and other major atmospheric gaseous components may influence the measurements.

The PA effect in gases can be divided into five main steps (Fig. 3.1):

1. Modulation of the laser radiation (either in amplitude or frequency) at a wavelength that overlaps with a spectral feature of the target species; an electrooptical modulation device may also be employed, or the laser beam is modulated directly by modulation of its power supply; the extremely narrowband emission of the laser allows the specific excitation of molecular states; the laser power should be modulated with a frequency in the range $\tau_{th} \gg 1/f \gg \tau_{nr}$, where τ_{th} is the thermal relaxation time, and τ_{nr} the nonradiative lifetime of the excited energy state of the molecule.

2. Excitation of a fraction of the ground-state molecular population of the target molecule by absorption of the incident laser radiation that is stored as vibrational-rotational energy; the amount of energy absorbed from the laser beam depends on the absorption coefficient, which is a function of pressure.

3. Energy exchange processes between vibrational levels (V-V: vibration to vibration transfer) and from vibrational states to rotational and translational degrees of freedom (V-R, T transfer); the energy which is absorbed by a vibrational-rotational transition is almost completely converted to the kinetic energy of the gas molecules by collisional de-excitation of the excited state; the efficiency of this conversion from deposited to translational energy depends on the pressure and internal energy level structure of the molecule; vibrational relaxation is usually so fast that it does not limit the sensitivity; however, notable anomalies occur in the case of diatomic molecules, such as CO, where vibrational relaxation is slow in the absence of a suitable collision partner, and of the dilute mixtures of CO_2 in N_2, where the vibrational energy is trapped in slowly relaxing vibrational states of N_2; the kinetic energy is then converted into periodic local heating at the modulation frequency.

4. Expansion and contraction of the gas in a closed volume that give rise to pressure variation which is an acoustic wave; the input of photon energy with correct timing leads to the formation of a standing acoustic wave in the resonator.
5. Monitoring the resulting acoustic waves with a microphone; the efficiency with which sound is transmitted to the microphone depends on the geometry of the cell and the thermodynamic properties of the buffer gas.

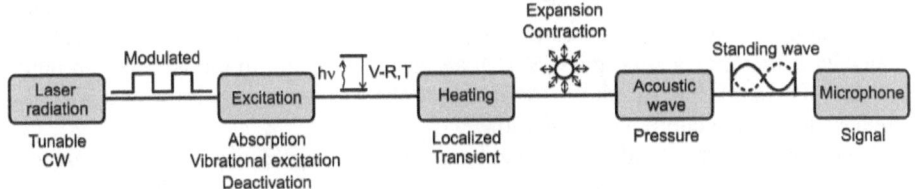

Fig. 3.1. Schematic of the physical processes occurring during optical excitation of molecules in photoacoustic spectroscopy.

The photoacoustic response is fast, and due to the high sensitivity in trace gas detection there is no need of large quantity of sample. Spectral dependence of absorption determines the nature of components, allowing simultaneous measurement of different gas concentrations in gas mixture. The proportional PA signal depends linearly on the absorption coefficient and gas concentration. The block diagram of the laser photoacoustic spectrometer is shown in Fig 3.2.

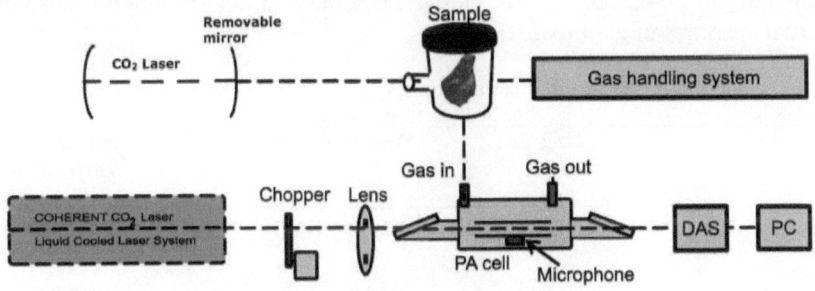

Fig. 3.2 - The CO_2 laser photoacoustic scheme.

Laser Photoacoustic Spectroscopy method (LPAS) is a very powerful investigation technique which is capable of measuring trace gas concentration at sub ppb level. LPAS experimental setup used in the present work consists of a CO_2 laser, a photoacoustic (PA) cell where the gas sample is analyzed, a vacuum/gas handling system, and a detection unit.

The CO_2 laser used in this work is a high power laser (50 W, in order to obtain a smaller detectable trace gas concentration) GEM Select 50 Laser System, Coherent, Inc., tunable between 9.2 and 10.8 μm on 73 different vibrational rotational lines. The kind and number of detectable substances is related to the spectral overlapping of the laser emission with the absorption bands of the trace gas molecules in the wavelength region 9-11 μm. Thus, the accessible wavelength range, tunability, and spectral resolution of the laser are of prime importance. If there is a complex mixture of gases, one should choose a proper absorption line for determination of concentration, where line overlapping with other gases is low [11-14]. The requirement for the gases to be detected is that they should possess high absorption strength in the wavelength range of the CO_2 laser. So, we chose for each gas the corresponding laser line with the highest absorption coefficient (see Table 3.1).

The laser beam is amplitude modulated by an optical chopper, focused by a ZnSe lens and introduced in the PA cell. The laser power used to excite the sample gas inside the PA cell is measured by a two channel powermeter. The acoustic waves produced in the PA cell are detected with four miniature microphones connected in series. The PA signal, proportional to the trace gas concentration is applied to a lock-in amplifier which detects and measures very small single frequency AC signals. The output signals of the lock-in amplifier and of the powermeter are then converted into digital signals and processed by a computer. A software program (TestPoint) for graphic and instrumentation permits to obtain and process the experimental results. The absolute trace gas concentrations are processed by the computer and the results are displayed on the screen. The gas handling system is an important part of the experimental set-up for the gas level concentration measurements, ensuring gas purity in the PA cell. The handling system can be used to introduce the sample gas in the PA cell at a controlled flow rate, to pump out the gas sample from the cell, and to monitor the total and partial pressures of gas mixtures.

Tabel 3.1. Type of gases measured from surgical smoke with the corresponding laser line and the absorption coefficient.

Gas	CO_2 laser line	Wavelength [μm]	Absorption coefficients [cm^{-1} atm^{-1}]
Acetonitrile	9P(16)	9.517	0.15
Acrolein	10R(14)	10.286	2.80
Ammonia	9R(30)	9.217	56.00
Benzene	9P(30)	9.636	2.00
Ethylene	10P(14)	10.529	30.40
Toluene	9P(28)	9.618	0.67

The PA cell is a longitudinal resonant cell is a cylinder with microphones located at the loop position of the first longitudinal mode (the maximum pressure amplitude). Some general considerations imply that the coherent photoacoustic background signal caused by window heating is decreased if the beam enters the cell at the pressure nodes of the resonance. The advantage of mounting the windows at the pressure nodes is well demonstrated, and the window heating signal is decreased by the Q factor. The laser beam enters and exits the cell at the Brewster angle.

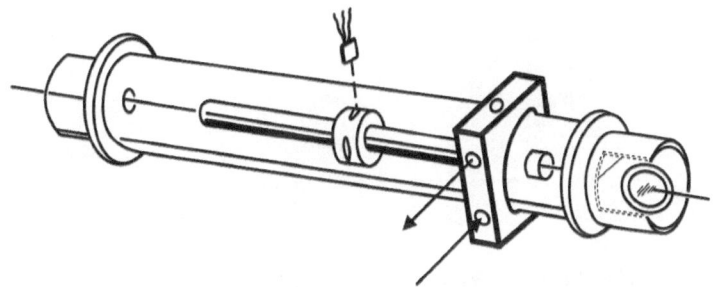

Fig. 3.3 Schematic of the PA cell designed for the first longitudinal resonance mode.

An important parameter of the PA cell is the responsivity R [Vcm/W] which is defined as the amplitude of the electric signal provided by the microphones on the unity absorbed power of the molecules on the unity length:

$$R = C \cdot S_M \quad (3.1)$$

Representation of this dependence is presented in Fig. 4.4, where for above measured average pressures $R \sim 300$ cmV/W and ~ 200 cmV/W respectively.
The minimum detectable concentration which can be detected with the LPAS system is calculated with the following equation:

$$c_{min} = \frac{V_N}{\alpha P_L C S_M} \quad (3.2)$$

where $V = V_N$ is the voltage of the photoacoustic signal for a signal to noise ratio equal with 1 (SNR = 1), α [cm^{-1}·atm^{-1}] is the absorption coefficient for a given laser line, P_L [W] is the unmodulated peak value of the power laser, C [Pa·cm/W] is cell constant, and S_M [mV/Pa] is the total responsivity of the four microphones (S_M = 80 mV/Pa).

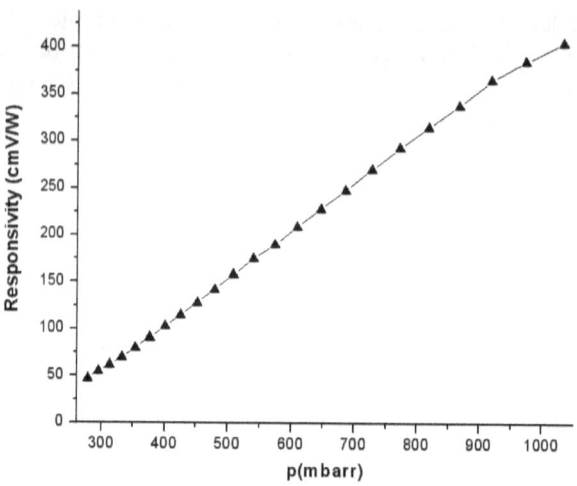

Fig. 3.4 The responsivity of the PA cell against the pressure.

To increase the accuracy of LPAS method for measurements of biomarkers in exhaled air of subjects, the sample gas is passed first through a trap filled with potassium hydroxide (KOH) pellets (with a volume larger than 100 cm^3) to remove the carbon dioxide and the water vapors before being collected into the PA cell and renewed after each measurement. Those measurements should not be altered by the molecules previously adsorbed on the PA cell wall and on the pathway, a cycle of N_2 washing was performed at the end of each sample. An average over several independent measurements at each line was used to improve the accuracy of the results.

3.3. Production and collection *in vitro* of surgical smoke samples

Laser smoke was produced *in vitro* by ablation of fresh porcine tissues placed in a glass cylinder [14] provided with a ZnSe window, which is positioned perpendicular to the laser beam, and two access pipes for handling main gases connected to a gas system and to the PA cell. The laser vaporization of the tissues was realized in a closed environment in nitrogen atmosphere and the smoke was collected for photoacoustic measurements in the PA cell. The smoke was introduced into the PA cell using nitrogen gas via a regulator valve to maintain a constant flow at atmospheric pressure.

The influence of the main components of surgical smoke, water vapors and carbon dioxide was removed by introducing before the PA cell a KOH trap. The pellets of the KOH trap were replaced before every measurement and the PA cell was cleaned after each measurement, first by using a vacuum pump and then the stream of nitrogen (purity 99.999%). The measurements correspond to a dilution of the smoke sample in 1 dm^3 of buffer gas (nitrogen).

The surgical smoke samples were analyzed without using a particle filter, so the results could be somehow altered by the presence of smoke particles inside the PA cell. Measurements were realized at the atmospheric pressure (1010-1034 mbar) and room temperature (23^0 C). As tissue samples we used pig skin, pig kidney, pig muscle, and pig heart that were stored at 4^0 C until the moment of use. The porcine tissues were obtained from a local slaughterhouse.

With our photoacoustic system we determined the concentration of acetonitrile, acrolein, ammonia, benzene, ethylene and toluene. We investigated different types of tissue at a vaporization laser power of 10 W and 15 W, respectively, and exposure times of 5 s and 15 s, respectively. For each tissue type (pig kidney, muscle, skin and heart) we performed several measurements and the presented results are the average of the determined gas concentrations (given in ppmV).

3.4 Results and discussions: quantitative analysis

The influence of laser power, exposure time and tissue type on the gas concentrations was investigated using the LPAS system. The surgical smoke was produced in vitro by CO_2 laser vaporization of porcine tissue in a closed environment. All gases (acetonitrile, acrolein, ammonia, benzene, ethylene and toluene) released from tissue samples were measured in the CO_2 laser wavelength range (9.2 – 10.8 µm).

The average gas concentrations measured in the smoke samples with our LPAS system are acetonitrile - 190 ppm, acrolein - 35 ppm, ammonia - 25 ppm, benzene 20 - ppm, ethylene - 0.410 ppm, and toluene - 45 ppm for kidney, muscle, skin and heart pig tissues, but gas concentrations may vary from sample to sample. The variations of gas concentrations for acetonitrile, acrolein, ammonia, benzene, ethylene and toluene with laser power, exposure time, and tissue type are presented in Fig. 4.5-4.7. We observe that the concentrations for all six gases increase with laser power and with exposure time and depend on the type of tissue sample. The higher concentrations were measured in pig skin sample. When pig heart is used as sample, we observed a different behavior of gas concentrations.

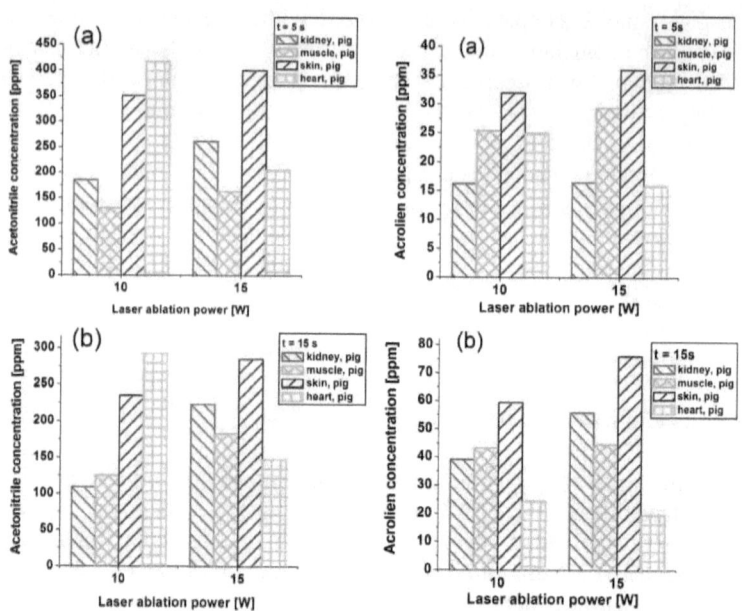

Fig. 3.5 - Gas concentration for acetonitrile (left) and acrolein (right) at P = 10 W and P = 15 W for: (a) exposure time of 5 s; (b) exposure time of 15 s.

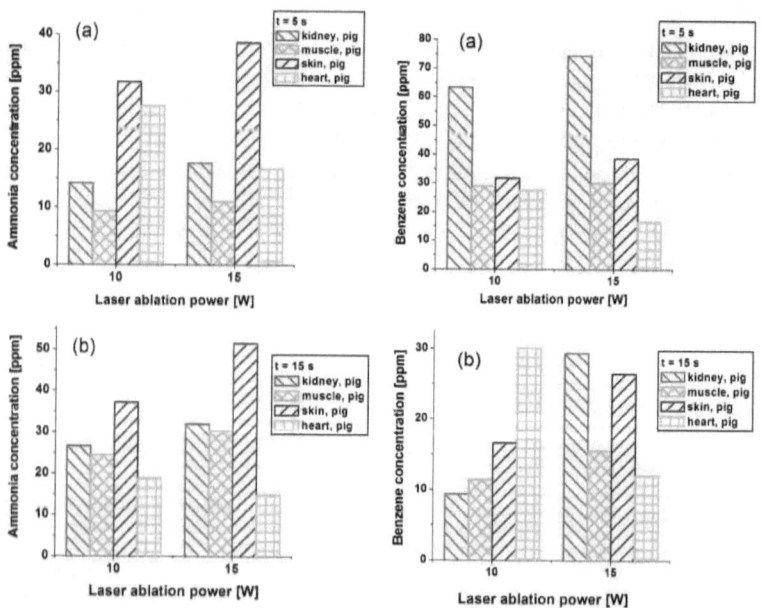

Fig. 3.6 - Ammonia (left) and benzene (right) concentrations at P = 10 W, P = 15 W: (a) exposure time of 5 s; (b) exposure time of 15 s.

In the case of tissue samples with a low percent of water, as pig skin sample, the gas concentrations (acrolein, benzene, and toluene) are greater than in the case of tissue samples with a high percent of water. This means that the humidity not only helps to estimate the role of the surrounding atmosphere, and reduces the concentration of hazardous chemicals, but it can be another way to reduce the emission of toxic substances during laser-tissue interaction. This result illustrates the role of water vaporization during laser irradiation of tissue. It is possible that during laser radiation interaction with a tissue that contains enough water; a water vapor cloud could be formed that screens the reaction of the surrounding atmosphere. The water content of the tissue influences the emission of volatile organic compounds, and this can be observed especially in the case of benzene and toluene.

The difference in concentration could also be due to scattering on soot particles, because the measurements were made without a particle filter.

The medical personnel is exposed over a greater period of time to surgical smoke and the surgeons who work at a distance of 20-40 cm from the area of smoke generation are exposed to the highest concentration of smoke. Standard surgical masks are inadequate to protect the medical personnel and the surgeon.

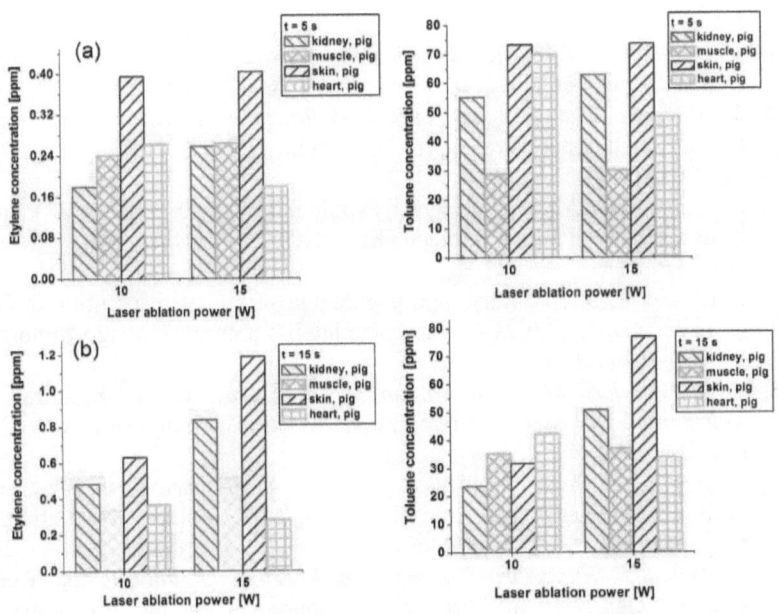

Fig. 3.7 - Gas concentration for ethylene (left) and toluene (right) at P = 10 W and P = 15 W for: (a) exposure time of 5 s; (b) exposure time of 15 s.

3.5 Conclusions

In the present study, a quantitative analysis of surgical smoke produced *in vitro* by irradiation of porcine tissues in a closed nitrogen atmosphere was made using a CO_2 laser photoacoustic spectroscopy system.

We demonstrated the presence of six toxic gases (acetronitrile, acrolein, ammonia, benzene, ethylene and toluene) in surgical smoke after laser vaporization at each measurement and the concentrations are of the order of ppmV.

These results demonstrate that the laser vaporization power and the exposure time are important parameters and gas concentrations are influenced by the water content of tissues. The toxic gas concentrations in smoke samples depend proportionally with the laser vaporization power and with the exposure time, but also depend on tissue type.

The results showed that the LPAS system proved its efficiency in analyzing a multicomponent gas mixture.

References

1. M. Gianella, M.W. Sigrist, Infrared spectroscopy on smoke produced by cauterization of animal tissue, Sensors, 10: 2694-2708 (2010).
2. M. Baggish, B. Poiesz, D. Joret, P. Williamson, and A. Rebai, Presence of Human Immunodeficiency Virus DNA in Laser Smoke, Lasers Surg. Med., 11: 197-203 (1991).
3. P. Hallmo, O. Naess, Laryngeal Papillomatosis with Papilloma Virus DNA Contracted by a Laser Surgeon, Em. Arch. Otorhinolaryngol, 248: 425-427 (1991).
4. L. Calero, T. Brusis, Laryngeal papillomatosis - First recognition in Germany as an occupational disease in an operating room nurse, Laryngo-rhino-otologie, 82: 790-793 (2003).
5. J. M. Garden, M. K. O'Banion, A. D. Bakus, C. Olson, Viral Disease Transmitted by Laser-Generated Plume (Aerosol), Arch. Dermatol., 138: 1303 (2002).
6. B. Ziegler, C. Thomas, T. Meier, R. Muller, T. Fliedner, L. Weber, Generation of Infectious Retrovirus Aerosol through Medical Laser Irradiation, Lasers Surg. Med., 22: 37-41 (1998).
7. D. Jewett, P. Heinsohn, C. Bennett, A. Rosen, C. Neuilly, Blood- Containing Aerosols Generated by Surgical Techniques - a Possible Infectious Hazard, American Industrial Hygiene Association Journal, 53: 228-231 (1992).
8. P. Heinsohn, D. Jewett, Exposure to Blood-Containing Aerosols in the Operating-Room - a Preliminary-Study, American Industrial Hygiene Association Journal, 54: 446-453 (1993).
9. K. J. Weld, S. Dryer, C. D. Ames, K. Cho, C. Hogan, M. Lee, P. Biswas, J. Landman, Analysis of Surgical Smoke Produced by Various Energy-Based

10. *Y. Tomita, S. Mihashi, K. Nagata, S. Ueda, M. Fujiki, M. Hirano, T. Hirohata*, Mutagenicity of Smoke Condensates Induced by CO_2- Laser Irradiation and Electrocauterization, Mutat. Res., 89 145-149 (1981).
11. *W.L. Barrett, S.M. Garber*, Surgical smoke-a review of the literature, Business Briefing: Global Surgery, p.1-7 (2004).
12. *D. Galbraith, S. A. Gross, D. Paustenbach*, Benzene and Human Health: A Historical Review and Appraisal of Associations with Various Diseases, Critical Reviews in Toxicology, 40: 1-46 (2010).
13. *L. Lehman-McKeeman*, Paracelsus and Formaldehyde: The Dose to the Target Organ Makes the Poison, Toxicological Sciences, 116(2): 361-363 (2010).
14. *M. T. Smith*, Advances in understanding benzene health effects and susceptibility, Ann. Rev. Pub. Health, 31:133-148 (2010).
15. *T. Fukushima, K. Abe, A. Nakagawa, Y. Osaki, N. Yoshida, Y. Yamane*, Chronic ethylene oxide poisoning in a factory manufacturing medical appliances, J. Soc. Occup. Med. 36:118–123 (1986).
16. *G. A. Ortolano, J. S. Cervia, F. P. Canonica*, Surgical Smoke, a concern for infection control practitioners, Managing Infection Control, p. 48-54 (2009).
17. *C.R. Yeh*, Surgical smoke plume: Principles and Function of Smoke, Aerosol, Gases and Smoke Evacuators, Surgical services management, 3(4): 41-45 (1997).
18. *D. C. Dumitras, A. M. Bratu, C. Popa*, CO_2 laser photoacoustic spectroscopy: I. Principles, p. 1-42, in CO_2 Laser – Optimisation and Application, p: 43-102, Editor Dan C. Dumitras, Intech (Croatia), ISBN 978-953-51-0351-6 (2012).
19. *A.M. Bratu, C.Popa, C. Matei, S. Banita, D.C.A. Dutu, D.C. Dumitras*, Removal of interfering gases in breath biomarker measurements, J. of Optoelectronics and Adv. Mat., 13(8): 1045-1050 (2011).
20. *C. Popa, D. C. A. Dutu, R. Cernat, C. Matei, A. M. Bratu, S. Banita, D. C. Dumitras*, Ethylene and ammonia traces measurements from the patients' breath with renal failure via LPAS method, Appl. Phys. B, 105(3): 669-674 (2011).
21. *C. Popa, A. M. Bratu, R. Cernat, D. C. A. Dutu, S. Banita and D. C. Dumitras*, Spectroscopic studies of ethylene and ammonia as biomarkers at patients with different medical disorders, U. P. B. Sci. Bull., Series A, 73: 167-174 (2011).
22. *C. Popa, A. M. Bratu, C. Matei, R. Cernat, A. Popescu and D. C. Dumitras*, Qualitative and quantitative determination of human biomarkers by laser photoacoustic spectroscopy methods, Laser Physics, 21: 1–7 (2011).
23. *D. C. Dumitras, D. C. Dutu, C. Matei, A. M. Magureanu, M. Petrus, C. Popa*, Laser photoacoustic spectroscopy: principles, instrumentation, and characterization, Journal of Optoelectronics and Advanced Materials, 9: 3655-3701 (2007).
24. *D. C. Dumitras, D.C. Dutu, C. Matei, A. M. Magureanu, M. Petrus, and C. Popa*, Improvement of a laser photoacoustic instrument for trace gas detection, U. P. B. Sci. Bull., Series A, 69:45-56 (2007).

25. D. C. Dumitras, D. C. Dutu, C. Matei, A. M. Magureanu, M. Petrus, C. Popa, M. Patachia, Measurements of ethylene concentration by laser photoacoustic techniques with applications at breath analysis, Romanian Reports in Physics, 60:593-602 (2008).
26. R. Cernat, C. Matei, A. M. Bratu, C. Popa, D. C. A. Dutu, M. Patachia, M. Petrus, S. Banita, and D.C. Dumitras, Laser photoacoustic spectroscopy method for measurements of trace gas concentration from human breath, Romanian Reports in Physics, 62: 617-623 (2010).
27. D. C. Dumitras, S. Banita, A. M. Bratu, R. Cernat, D. C. A. Dutu, C. Matei, M. Patachia, M. Petrus, C. Popa, Ultrasensitive CO_2 laser photoacoustic system, Infrared Physics & Technology Journal, 53: 308-314 (2010).
28. D. C. Dumitras, D. C. Dutu, C. Matei, R. Cernat, S. Banita, M. Patachia, A. M. Bratu, M. Petrus, C. Popa, Evaluation of ammonia absorption coefficients by photoacoustic spectroscopy for detection of ammonia levels in human breath, Laser Physics, 21: 796-800 (2011).
29. M. Petrus, C. Matei, M. Patachia, D.C. Dumitras, Quantitative *in vitro* analysis of surgical smoke by laser photocoustic spectroscopy, Journal of Optoelectronics and Advanced Materials, 14 (7-8):664-670 (2012).

I want morebooks!

Buy your books fast and straightforward online - at one of the world's fastest growing online book stores! Environmentally sound due to Print-on-Demand technologies.

Buy your books online at
www.get-morebooks.com

Kaufen Sie Ihre Bücher schnell und unkompliziert online – auf einer der am schnellsten wachsenden Buchhandelsplattformen weltweit! Dank Print-On-Demand umwelt- und ressourcenschonend produziert.

Bücher schneller online kaufen
www.morebooks.de

OmniScriptum Marketing DEU GmbH
Heinrich-Böcking-Str. 6-8
D - 66121 Saarbrücken

Telefax: +49 681 93 81 567-9

info@omniscriptum.de
www.omniscriptum.de

www.ingramcontent.com/pod-product-compliance
Lightning Source LLC
Chambersburg PA
CBHW031547210526
45464CB00003B/1190